WITHDRAWN FROM
THE NEELB LIBRARY SERVICE
ON
FOR SALE AT

The sun

Georg Blattmann

The sun
The ancient mysteries
and a new physics

Floris Books
Anthroposophic Press

Translated by Robin Alexander

First published in German under the title
Die Sonne – Gestirn und Gottheit
by Verlag Urachhaus, Stuttgart, 1972.
First published in English in 1985 by Floris Books.

© 1972 Verlag Utrachhaus Johannes Mayer GmbH & Co KG Stuttgart
This translation © Floris Books, Edinburgh, 1985.
All rights reserved. No part of this publication may be
reproduced without the prior permission of the publishers.
Published in the United Kingdom by Floris Books,
21 Napier Road, Edinburgh EH10 5AZ.
Published in the United States by Anthroposophic Press, Inc,
258 Hungry Hollow Road, Spring Valley, NY 10977.

British Library Cataloguing in Publication Data

Blattman, Georg
The sun : the ancient mysteries and a new physics.
1. Religion and science—1946–   2. Sun—
Religious aspects
I. Title   II. Die Sonne. *English*
261.5'5   BL253

ISBN 0-86315-029-2 (Floris Books)

ISBN 0-88010-148-2 (Anthroposophic Press)

Printed in Great Britain
by Billing & Sons Ltd, Worcester

Contents

The objective marked out   7

1   The Wholly Other

1.1	The winged sun	17
1.2	The corona	21
1.3	Simplicity needs courage	24
1.4	Where is the sun?	27
1.5	How big is the sun?	31
1.6	Mysterious twilight	35
1.7	Mother and father sun	42
1.8	Sun hands	44
1.9	Inside the sun	50
1.10	More about the corona	53
1.11	What do we mean by hot?	55
1.12	'That which is inside is also outside'	60
1.13	The hollow sun	63
1.14	Floating pearls	66
1.15	Cosmic whirlpool	68
1.16	The sunbird descends	73
1.17	Pre-Christian vision of Christ	77
1.18	Expectation and fulfilment	84
1.19	The sun guardian as guest	90
1.20	The sun as procreator	94
1.21	The threefold trial	98
1.22	Cosmic consciousness	102
1.23	The sun path	105

2   Contrary Worlds

2.1	The opposing players	115
2.2	Man as mediator	119

2.3	The sun-origin of love	123
2.4	More contrariety	128
2.5	The being outside itself	130
2.6	Space and counter-space	134
2.7	An excursion into geometry	139
2.8	Kernel and husk	151
2.9	The case of the pickled gherkin	154
2.10	The plant between sun and earth	157
2.11	Turn-about on the sun's rim	160
2.12	Eighty-eight	164
2.13	The sacred loop	170
2.14	Christ in you	177

3   The Kingdom of Light

3.1	Sunspots	185
3.2	Symptoms on the skin	190
3.3	*Qui Tollit Peccatum Mundi*	193
3.4	Who is the sun-king?	200
3.5	New discoveries in the social sphere	205
3.6	The hierarchy as an order	210
3.7	My kingdom is not of this world	216
3.8	Sun God and brother of man	223

Epilogue: Limitations and gratitude	227
Acknowledgements	233
Bibliography	235
Index	237

References

The system used in this book cites author and year of publication, followed by volume (if necessary) and page. The full title and publication details are in the bibliography.

The objective marked out

Everyone knows it. Everyone loves it. Everyone longs for its presence when it hesitates to appear, because darkness—spatially rolled up in heaps of cloud, temporally shortening the day's length—inhibits its radiant appearance.

Every heart brightens in its brightness. Joy rises when it rises. We all live with it, through it. Its absence is our loss. Its presence our joy and comfort. It is quite close to us: *the sun*, the parent star of our cosmic home. But what do we really know about it?

When we ask for the results of astronomical research we learn that an almost incalculable number of details is known. However, as we study this information, a peculiar feeling comes over us. The more scientists tell us about their research into the sun and the better we get to know their theories, the more estranged we become from the nature of the sun as we know it in our daily experience. It is characteristic of the theories of astrophysics that they alienate from normal experience those who engage in them. Can we, then, reconcile the statement that the sun is a powerful nuclear reactor with the invigoration which a sunny summer's day brings us? Can that which in our experience is the most deadly in the world — the destructive power of the nuclear process — be seen as one with the universal giver of life? Such an attempt drives one into a kind of schizophrenia.

But must the actual experience of scientists lead to such absurd results? Perhaps there is another way of evaluating the observed phenomena? If it were possible simply to let the perceptible phenomena speak for themselves then presumably there ought not to arise any contradiction between daily experience and the work of scientists.

Here, in the first instance, is an objective for our investigation. We wish to develop solar research afresh from basic principles; allow the observations to have their say without thereby alienating ourselves from common understanding. Thus we shall get a picture of the sun which, in conformity with immediate natural experience, satisfies equally our understanding, our feelings and our need for fact.

The second objective of our investigation can be be explained from the point of view of psychology.

The sun-complex

Whenever our conscious attitudes oppose subconscious ones, listlessness and frustration and eventually illness, may arise. This is known to psychology and is familiar to almost everyone.

To begin with, let us consider the 'repressed complexes'. These are mental images, desires and feelings which, for some reason or other, are painful to our waking consciousness and are therefore swept away into the abyss of forgetfulness. They don't lie quietly there, however, but lead a restless life of their own and in this way become the source of illness. When a skilful doctor or pastor succeeds in eradicating these unconscious pockets by helping the patient to raise the inhibitions once again into the light of consciousness where they can be inwardly worked over and the knots loosened, then a serious illness can often be surprisingly quickly cured by 'spiritual' means or, as one says, through 'psychotherapy'.

In our investigation of the subconscious, however, we come up against yet another phenomenon which has a bearing on our problem. In the subconscious there are images and facts of knowledge which could not have been repressed within our lifetime but which are present 'from the beginning'. Such primeval elements are to be met with throughout humanity, transcending the boundaries of nation and race. For this the concept 'collective unconscious' has been coined. This means that all human souls have a common basic possession, the supporting foundation of our whole inner life, which we must reckon with. Just as it can be dangerous not to consider the nature of the foundations when making building alterations, so likewise when the development of daily conscious events contradicts the hidden foundations of the house of life, then ill-feeling, disharmony and finally nervous breakdown ensues.

C. G. Jung, the brilliant psychologist through whose research the 'collective unconscious' was discovered, formulates the consequences of its activity as follows (1957, 7:70):

> This is the reason why men have always needed demons and cannot live without gods, except for a few particularly clever specimens of *homo occidentalis* who lived yesterday or the day before, supermen whose 'god is dead' because they themselves have become gods — but tin gods, with thick skulls and cold hearts.

The 'gods' and the 'demons' belong to the fundamental dimensions of the human soul. When, in earlier times, people narrated their great myths of the gods, a valuable spiritual therapy was thereby exercised on human beings from childhood on. The unconscious dimensions of the common spiritual life were brought into the light of consciousness and were thereby deprived of the disease-inflicting power they give rise to so long as they operate hidden and unknown in the depths of the soul. Humanity must fall sick when it does not receive a knowledge of the gods commensurate with subconscious reality.

The relationship of sun and Christ belongs to the basic constitution of our common subconscious soul life. Each person carries this fact in the hidden depths of his being. But he does not know it.

The peoples of earlier times (and present-day primitive peoples wherever they are able to live undisturbed) knew how to bring this basic element of the human soul into the light of conscious life, namely, through practice of their holy ritual, through the continual repetition of their sagas in which they honoured the sublime divine being of the sun in word and action, responded lovingly to its inclination towards the earth and prepared themselves for its 'coming'.

We can no longer tread the old paths of the sun religions. Indeed, through religious and scientific education they have been made despicable to us and expelled from our understanding. Our wonderful knowledge of the sun-god is wiped out. The mystery lies motionless, sterile and dead at the bottom of the psychic ocean. But things in the psychic realm differ from those in the sphere of nature. Though a rock can rest peacefully on the bed of the ocean without causing harm, that which goes unnoticed and unlived in the soul becomes a source of disturbance. We all bear the sun-complex within us; and it disturbs our mental health. An injurious influence makes itself doubly felt when its reality is ignored.

Christians talk about Christ. They think they know him — and yet they have lost sight of his kinship with the sun. Thus, as time passes, they fail to find true satisfaction in their convictions because, without their being aware of it, a large portion of reality is missing. The submerged complex remains unraised. Conventional theology is, therefore, superficial and must necessarily remain so.

On the other hand those who carry out research into the nature of the sun, the astrophysicists, are even more alienated from insight into the primal connection between sun and Christ. Therefore, neither can their assertions satisfy us. Rather do their thought structures settle on our minds like a systematized obsession. They stifle souls by crippling the free emergence of primal memories in which the divine vitality of the sun's heavenly body might still be intuitively contemplated.

In this matter, be it noted, only the theories and theorems which scientists have constructed from their observations are being put in the pillory. What they have really seen with the naked eye and with the telescope is something we ourselves shall make full use of in what follows.

Sun longing

Genuine sun lore — which raises the content of the subconscious in its dual aspect to awareness — meets today an unusual echo which demonstrates how sensitive is the old scar in the soul left by the erosion of the sun-god mystery.

It happened for example at a youth conference that conversation turned casually to the subject of the sun and mention was made of its true nature as developed from the astronomical facts of observation put forward in this book; immediately, a hectic curiosity broke out among the young people. Different as were their scholastic and social backgrounds (they came from seven towns in different provinces and from all types of school) nevertheless, they all, of one accord, showed themselves to be suddenly appealed to and aroused. Independently of whether they were scientifically or artistically gifted, ill-behaved or childish, the sun theme gripped them. This was variously expressed. The reactions ranged from a vehement opposition to an insistent thirst for knowledge; it was typical of the mode of behaviour expressed in situations where deep levels of the psyche are touched.

This example is representative of many others. Numerous descriptions, both by word of mouth and in writing, bring home to us how people, ranging from primary school children to the inmates of old people's homes, have again and again the same experience. Whenever one speaks of sun lore in conjunction with the facts of Christ's Life then an integration of universe and subconscious mind is spontaneously established in the soul's depths, and, in blissful moments, gives rise to harmonious sincerity.

Synthesis

If it were possible to outline a true picture of the sun and to combine it with a genuinely modern understanding of Christ, then such a synthesis of science and religion would radiate a fundamentally curative influence on the disrupted spiritual life of present-day humanity. To this end an unusual exertion would have to be mustered, both in science and in religion. But it would be worth the effort. It is a question of providing basic psychotherapeutic and pastoral care for the consciousness of our humanity and thereby for life on earth.

To attain the objective we have in mind, we shall make a threefold approach.

In the *first* section, we shall clarify a group of genuine observable phenomena in the heavenly body of the sun. We shall thus be led to a conclusion that is quite new and unusual *vis-à-vis* the prevailing conception. The sun appears, as it were, 'against the light'. It presents itself as 'the Wholly Other'. It challenges us to modify our concepts. And in the extended field of vision, there emerges unexpectedly a spiritual figure which seems to stand in close relationship to the sun's being. At this, the first stage, it will be possible to behold a 'Divine-Other', universally great and blessed by the sun.

What has thus been gained will be augmented in the *second* section by an exact mathematical description. On the basis of the contrast between sun and earth much will be clarified and explained. The 'counterpoised worlds' take on individual identity and throw light on one another. In the interplay of polarities, the relationship of the living powers of the Christ-being to natural and spiritual events will be understood on a new level.

Thirdly, consideration of the sun spots will lead us more deeply into the mystery of the sun. Everything we have learnt will be gathered together into a general impression by the apparently special peculiarities of sunspot activity. 'The kingdom of light' can be experienced as a unity. Ebbing and flowing, it draws near to us, into the passing of our earthly life. Thus, as a formative spiritual reality, we realize the third stage of our own future.

1

The Wholly Other

O Sun, the face of truth is veiled by thy golden disc. Withdraw it, that I, a seeker after truth, may behold Truth's splendour.
                                        Isha Upanishad

1.1 The winged sun

Does the sun have wings?

Though at first the question may surprise us, nevertheless it somehow throws a spell on us. Let us for once allow ourselves to remain spellbound by this reflection, and try to track down the mystery of the winged sun.

The sun with wings was a thoroughly familiar sight to an ancient Egyptian. He knew from a great number of pictures on the temple walls and in tombs, how the sun-god ought to be worshipped and depicted, namely, with the widely-spread wings that carry the shining disc across the heavens.

That the sun has wings presents no problems to children of primary school age. (One can say, for instance. 'We don't generally think about it, because our eyes are dazzled by the brilliant light and therefore we cannot easily look at the sun.') When we mention this fact to them, it is as if something they had long known, but forgotten, was called back into memory. Somehow or other, it is familiar and obvious to them. That other types of enlightenment are given in school or on television does not affect their receptivity. But adults, too, show a ready willingness in this century to look beyond school wisdom and the mass media. Therefore, though it may surprise, it shocks no-one to see the winged sun turn up again in the subway on the stand of young pop artists. There may be a row of pendants for sale, worked up in cheap metal, depicting a sun between two wings. Upon closer

Figure 1. Winged sun on the back-rest of King Tutankhamun's throne.

inspection one recognizes, attached to the wings, a delicate female figure, kneeling, and holding the gleaming disc in her raised hands. For all that, what matters is the uniform image of sun and wings.

Is it only a fashionable brainwave that brings this venerated image as a gimmick on to the market? Or has the manufacturer come up against a subliminal need of human souls when, in this

1.1 THE WINGED SUN

day and age, he finds a market for the ancient symbol? Does not the remarkable encounter with the younger generation in the underground passages where they have their rendezvous, speak clearly of a profound longing in human beings for sun lore? A longing that is so powerful that it readily grasps anything that brings freedom from the fatally impoverished knowledge of today.

Whether we follow the trail of Egyptian wisdom, or strive as teachers to keep an open mind for unusual truths, or whether we try to make psychological diagnoses of daily experiences, we always come up against the one basic truth: In the depths of the human soul the image of the winged sun survives in purity and in beauty, waiting only to be awakened.

However, before we venture to explore this feature, let us hear what science has to say about it. As students of cultural history we need to know what truth there is in the images of ancient religions. As educators of children we have the responsibility of ensuring that what we teach can withstand the rigour of precise thinking. As human beings of our time, we demand inside information on how the archetypes of our soul life tally with the cold facts of worldly knowledge.

Let us, therefore, ask the specialist, What does the science of astronomy have to say about the reality of the winged sun?

The answer is astonishing. The picture of the sun with wings crops up time and time again in the investigations of the astronomer. So often, in fact, that he does not consider it anything remarkable. Everyone appreciates that it is well worth seeing. And no-one giving a lecture on the sun will miss the opportunity of delighting his audience with a picture. Nor is it absent from any textbook on astronomy, or from any popular writing dealing with the sun theme. As we survey the literature on astronomy this is how we continually see it sketched and photographed in numerous variations, and every time we see this wonderful picture our enthusiasm is kindled afresh. Only, there is simply no-one today who would describe what he saw as a 'winged sun'. Many an

THE SUN

Figure 2. Sun's corona on February 25, 1952.

astronomer reading these lines will find this use of words strange, though the words do suggest themselves to an impartial observer. Admittedly there is one reservation. Where the layman expects to see the sun's disc in the middle we see only a yawning black hole. However, this peculiarity of the modern picture compared with the depiction of the ancients has its own testimonial value. We shall return to this later.

The specialist uses a particular designation for the plumed pencil of rays at either side of the darkened sun disc, a name that has its own charm and captures the spell of the phenomenon. He calls it the sun's 'corona'.

1.2 The corona

The Latin word for wreath or crown, *corona* is a very appropriate name for the exceptional beauty of this phenomenon which has fascinated astronomers for so long. The uniqueness of the spectacle is heightened by the awareness that it is only very seldom that a view of the sun's corona is possible. In the past hundred years, that is, since systematic observations were made and recorded, vigilant scientists have altogether only been able to reap about a hundred precious minutes' viewing. And this is not because they were lacking in ability. On the contrary, those who desired so ardently to observe the phenomenon, spared neither effort nor finance in order to exploit each second of the scanty minutes granted them. They had to contend with a heavenly occurrence whose behaviour towards human beings was indeed shy. Only in the short space of time of a total eclipse of the sun can one catch a glimpse of the matt-white, almost metallic lustre of the corona, stretching its plumed arms far out into space.

[The] silver grey rays which suddenly shoot forth from behind the rim of the moon as total eclipse takes place is described enthusiastically by the few who are able to enjoy the drama of such an eclipse, as one of the most wonderful sights . . . The corona appears as a bright aura of rays surrounding the darkened sun and like everything on the sun, their shape and extent change greatly, but the changes can only be compared through the great intervals of

THE SUN

Figure 3. The fibrous cluster of rays in the corona change shape as time passes.

time between one total eclipse and the next; during the few minutes of the actual eclipse the projected rays, often very long, retain their shapes. These are sometimes quite straight, sometimes curved in a characteristic manner like iron filings round the poles of a magnet. (Meyer 1897)

The outer contour of the corona exhibits a distinct change with the sun spot cycle. The corona at the time of sunspot maximum is fairly symmetrically distributed round the whole sun with rays on all sides and really gives the impression of a wreath of sunbeams. The corona at sunspot minimum, on the other hand, possesses a wide extension within narrow heliographic latitudes about parallel to the equator,

1.2 THE CORONA

while at the poles it is much weaker . . . On a small scale, apart from occasional bright knots, the corona does not show any actual structure, but a great number of arcs and rays. (Stumpff 1957)

By means of an ingenious optical device, the coronagraph, it has also become possible recently to acquire information about the corona outside the times of eclipse. However, the results achieved with this instrument lag far behind those gained by observations during a natural eclipse. For this reason, expeditions are still fitted out and sent off to distant parts of the world, to the South Seas, to central Asia and even to the polar regions should a total eclipse of the sun be expected there. On no account must an opportunity of seeing this mysterious wreath of sunbeams be missed.

Indeed, the enthusiasm inspired by the sight of the corona and the fascination that draws solar scientists ever and again to observe it must rest in the fact that in this plumed crown with its far-reaching, radiant veil, something of the true being of the sun is expressed. Is it not symbolic that, at the very moment when the corona becomes visible, the sun's disc disappears? Our attention is drawn away from the brilliant 'ball' in the middle and directed to the circumference. Is not this denial imposed upon us by natural law an indication of a hitherto little heeded but fundamental truth about the sun? At this point we must turn our consideration to the most simple elements of our experience.

1.3 Simplicity needs courage

If we were to ask what our first impression of the sun is, we should probably say, 'When the sun rises, the whole sky is light'.

This fact is so simple that we hardly dare put it forward as a matter for serious consideration. And yet in saying this we are expressing a characteristic of the sun which distinguishes it from all other heavenly bodies. The moon and stars have their own limited light which makes them stand out against the dark sky. The sun, on the other hand, becomes visible with such omnipotence that it determines the whole of its surroundings and extinguishes the background. It lies in the nature of the sun that its appearance is not confined to a limited space but is characterized by a supra-spatial omnipotence.

We must now have the courage to assess these simple perceptions on their true merit. Today we have a tendency not to trust our sense perceptions. Over-awed by the authority of science, which has introduced the spectral notion of 'sense-deception,' we have learned to be suspicious of and to belittle the experience of our senses. We constantly defer to the findings of scientific research divulged by instruments and calculations.

But is it not so with all knowledge that what is difficult must rest on what is simple? The natural order of things would be turned upside down were we to proceed otherwise. No wonder we get unnatural results. Standing on one's head is not good for the attainment of knowledge. One cannot make the first step

1.3 SIMPLICITY NEEDS COURAGE

depend upon the last. Rather another step must follow the first; only thus can a healthy progress be made. Infringement of this rule will be requited in deadly earnest sooner or later. Indeed, a century of scientific enterprise has distempered our thinking and behaviour and is proving deadly. In his far-reaching reflections on the present-day situation, the atomic physicist Walter Heitler (1963, 3) says: 'It is so much easier to damage and destroy life by the aid of science than to create life that — so it would appear — something of this tendency must be inherent in present-day science itself.' And in his excellent paper, he clearly illustrates the impulses which give rise to destructive tendencies.

The absurdity has arisen because the top has been laid at the bottom. Instead of starting from the immediate observations of a healthy understanding, upon which the more advanced and complicated structure could be built, we have as starting point abstract statements derived from scientific apparatus and involved calculations, and these are accepted to the detriment of our own perceptions, and the pure joy of observing suffers in consequence.

It is now clear to many that the *modus operandi* of the sciences is in need of reform. Did not Einstein and several of his outstanding colleagues urge a new form of thinking! Keen sense perception is the only sure foundation upon which a life-promoting knowledge compatible with human dignity can grow. Therefore, everyone, even the most unsophisticated, who has the courage to trust his senses, can say that he is doing curative work towards overcoming a blunder in our present mode of living. The consequences of such a reorientation will become strikingly evident in our essay on sun research.

The activity of modern science as expressed in physics, has the nuclear reactor as its inevitable end result. In such a machine 'the events taking place contradict life more than any other known physical process' states the Basle physicist, Professor M. Thurkauf.

A totally different picture would emerge, however, if those

elements of experience which everyone can vouch for were taken as the starting point of knowledge. But to start in such simplicity needs courage. It will become clear to us that the sun is the source of life in the universe and as such is the perfectly visible equivalent of a divine spirit. All the specialist knowledge which we obtain through observations made by precision instruments and careful calculation can be contained by this true representation of its being.

After these reflections on his own aptitude, the reader will be able to return with renewed joy to the contemplation of the sun.

1.4 Where is the sun?

The simple phenomenon which we have summed up in the words, 'When the sun rises, the whole sky is light', is most evident when it is cloudy. Under these conditions, the sun is experienced as flat white light, embracing the whole vault of the sky. The thicker the clouds the more telling the experience. Thus it is just when the sun's disc is covered that the singularity of the sun is most discernible. Again, we learn something new about the nature of the sun when the clouds have moved on and the sky is a deep transparent blue.

Where, then, is the sun? First, let us ask a small child. We assume that she, at least, will be unbiased. Perhaps she is standing in the middle of a sunny meadow among the gently swaying grass and flowers. Bees are humming. Otherwise, it is quite still. 'Where is the dear sun?' Whereupon, she raises her little arms. Her eyes sparkle with joy as she looks around her and stretches her hands out in all directions like feelers in the warm afternoon air, 'There!' The child is bathed in this 'there' of overflowing sunlight. 'There' — that is to say, in everything that surrounds me. Does that not remind us of the words of the great philosopher Schelling? 'The sun does not only shine where it is, it is also there where it shines.' Thus does a trained thinker formulate what he discerns, in delicately chosen words. What he means is in reality no different from what the child is expressing when she calls out, 'there' and waves her arms around. The sun is not in some remote, empty space, inaccessible to human beings. *It is where it shines*, namely, here among the flowers, the animals and people, bathing and penetrating everything with its abundant rays of light.

Had we asked an adult instead of a child, we would have received a different kind of answer. But an answer no less significant as we shall see. When one asks someone on a bright clear day, 'Where is the sun?' the chances are that he will say a little petulantly, 'Can't you see for yourself, then?' If one persists in seeming the fool, he will probably stretch out his arm and point at the bright spot in the sky. But then we notice something. At the same time as he points out the sun in the sky, he jerks his eyes away from it, blinking and blinded. Where is the sun now? 'There!' says our informant with a gesture; but he cannot point to this 'there', without at the same time shielding his eyes.

Again the words of an eminent man who wrote down his experience of light confirms our own observation, 'There was no use my seeing the sun high up in the sky in its place at noon, since I was always searching for it elsewhere. I looked for it in the flickering of its beams, in the echo . . .' wrote Jaques Lusseyran (1963, 5). This man, who went blind in childhood, immersed himself lovingly in light with a fervour seldom shown by those who can see. He lost the sight of both eyes in an accident when he was eight. And as though he foresaw his coming blindness, he felt those previous years to have been dedicated in childish ardour to the light. 'I was eating sun . . .'

After the operation, his dialogue with light continued inwardly. And it became a continuous exercise for the blind man in his further life. It is not merely a reminiscence. What he remembered took shape in his mind as new, profounder truth. The sun is born again in him, in its true form: 'I liked seeing that the light came from nowhere in particular, but was an element just like air. We never ask ourselves where air comes from, for it is there and we are alive. With the sun it is the same thing.'

If we now put what we have discovered together with what

1.4 WHERE IS THE SUN?

others have stated along the same lines, then we can listen to the sun's own being as it speaks to human experience: 'You think you will look at me, you think you will fix your gaze on me, but I won't have it! I am not what you think I am. If you can't bear to look at the dazzling glow of the disc which burns down on you from the sky, then there is a good reason. It is not in that locality that you will come to grips with my being. I am not to be identified with that blinding sphere which makes your eyes flinch when you seek me there. As your eyes turn away, so let also your thoughts turn away from that place, and move out into the surroundings. I am not a being of the here or there. You shall not say of me, "Look, here it is or there it is." You shall know that I am omnipresent.'

A criticism of the above has yet to be met. Someone could object that at sunset one can very well look at the ball for a period of time and fix its sharply defined shape on the horizon. But whoever refers to this phenomenon must also ask, has not the red beacon yonder on the horizon been robbed, for the most part, of its essentially sunlike nature? In coming 'down to earth' it has suffered loss and appears in the heaviness of dusk only as a shadow of its true self. Dim and without its warming rays, the sun disc appears as a side effect whose weak constitution allows of no weighty deductions.

We are heading in the right direction if we base our thesis on our experience of the midday sun, when it is exerting its greatest power.

The question, 'Where is the sun?' is answerable in one word: *everywhere*. The stunning simplicity of this formula should not let us overlook the profound consequences that follow from it. There is all the difference in the world between having to do with an object situated millions of miles out in space, remote from earth, and with a being whose chief characteristic is its omnipresence and its taking both ourselves and our planet into its universal embrace.

However, what does the science of astronomy have to say to such far-reaching conclusions drawn from everyday experience? We shall see. When we now consider the results of scientific research, hitherto holding second place in our enquiry, we discover that they are in full agreement with our findings.

1.5 How big is the sun?

When I ask children how big the sun is, they are perplexed at first and remain silent. Adults would, no doubt, react in the same manner; unless it chanced that one of them was an expert who could rattle off numerical data on request, and promptly supply the information, 'one million kilometres in diameter.' Why should the layman burden his mind with such things? No-one blames himself if he cannot answer these questions.

And yet, there is an answer which anyone can give, no matter how poor he is at figures. A child that has once learnt it will never forget, and furthermore, he will acquire a strong sense of reality — though this answer contains a very much greater, indeed a huge, spatial dimension. What is this answer? We must work our way step by step towards it, just as research itself only gradually arrived at this insight.

Let us recall the image of the corona. Astrophysicists say that a photograph falls far short of presenting us with what the observer actually sees. (That is why big expeditions still take place.) In the short period of total eclipse, the human eye is slow to adjust to the weak light. No sooner does it begin to discern those faint extended pencil rays — too faint to be recorded on film — than it is blinded by the blazing light breaking through from the sun's disc. Nevertheless, the spectator receives an indelible impression. This is not mere enthusiasm for the beauty of what he has seen, but a well-founded suspicion that the sun's wreath of rays

THE SUN

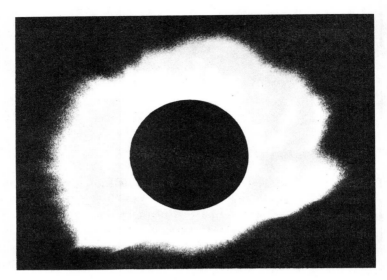

Figure 4. The field of the visible corona spreads out to a great distance. How far do the rays really reach?

stretches much further afield than is apparent in the short span of time it is visible.

Many facets of scientific research would need to be combined in order to show that the veil of the corona spreads out into infinite space and threads its way through the depths of the planetary system.

Scientists differ over this in the boldness of their statements. Since we have all been brought up on the Copernican view of the solar system, it is easier to give dimensions in terms of this scheme. One astronomer will say that the corona's rays run out just beyond the orbit of the inner planets, others see them as stretching just beyond the earth's orbit, whilst yet others infer that they stretch

1.5 HOW BIG IS THE SUN?

NICOLAI COPERNICI

net, in quo terram cum orbe lunari tanquam epicyclo contineri diximus. Quinto loco Venus nono mense reducitur. Sextum deniq; locum Mercurius tenet, octuaginta dierum spacio circũ currens. In medio uero omnium residet Sol. Quis enim in hoc

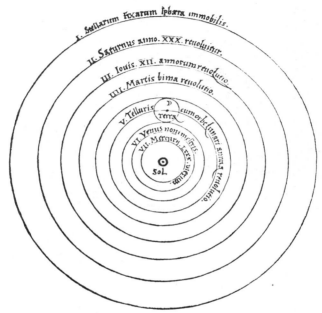

pulcherimo templo lampadem hanc in alio uel meliori loco po neret, quàm unde totum simul possit illuminare? Siquidem non inepte quidam lucernam mundi, alij mentem, alij rectorem uo= cant. Trimegistus uisibilem Deum, Sophoclis Electra intuentẽ omnia. Ita profecto tanquam in solio regali Sol residens circum agentem gubernat Astrorum familiam. Tellus quoq; minime fraudatur lunari ministerio, sed ut Aristoteles de animalibus ait, maximā Luna cū terra cognationẽ habet. Concipit interea à Sole terra, & impregnatur annuo partu. Inuenimus igitur sub
hac

Figure 5. The planetary system of Copernicus. Woodcut. 1543.

much further and even reach the planets Mars, Jupiter and Saturn. Does that bring us to the end of our imagination? Not at all. We can well conceive it possible that the most distant planets are embraced by the corona's arms. Admittedly, when we travel so far, we are using the expression 'corona' in a somewhat unusually extended sense. However, it is not a question of a choice of words but of a useful conception of reality. At any rate, science is moving towards the view that the corona stretches out over the whole planetary system. Our most recent indications have been obtained with the aid of satellites and space capsules.

In answer to the question of the size of the sun, we have now ascertained that, in its diffusion, this central star is coextensive with the bodies that belong to it. *The sun is as big as the planetary system.*

From this simple proposition which no reader will find difficult to grasp, valuable knowledge can be deduced. It will be clear, at this point, that we are talking about 'size' in terms that allow no comparison. Whoever talks about the sun must give up the customary attitude to celestial bodies as graspable objects, as huge spheres. The sun is different. The nature of its being is of a higher order. That is to say, the scene is laid on a plane of existence which is beyond comparison with anything else. The sun cannot be put on a par with other members of the planetary system. Our understanding cannot encompass it, because it encompasses absolutely everything. We must develop an understanding 'from within' to grasp the inner being of the sun.

1.6 Mysterious twilight

One can well ask: Do the radically new conceptions of sun and planetary system, advanced in the last chapter, rest on the hunches of a few astronomers and, at best, measurements made from satellites? This would run contrary to our principle that what a person sees and witnesses has priority. Assuredly, we have no right to be satisfied with the foregoing. There is, however, a more direct way of gaining knowledge about the sun's all-embracing dimensions which we shall now develop.

For a long time, observant students of the night sky have noticed a mysterious light, which because of its weak illumination is not always easy to see. The conditions are most favourable in spring and autumn. The observer in the tropics gets a much better view of this phenomenon than is ever possible in temperate latitudes because of the different twilight conditions in the neighbourhood of the equator.

Soon after sunset, or shortly before dawn, a delicate triangle of light, apex uppermost, rises above the horizon. It is about as bright as the Milky Way. However, upon closer inspection this enchanting shimmer is seen to be a homogeneous mistlike brightness, unlike the Milky Way which is a profusion of individual points of light. Under particularly favourable circumstances, one can observe a second, similar cone of light, like a weakened counter-image, in the opposite direction. Sometimes we can even see a luminous trail which, stretching over the sphere of

Figure 6. The zodiacal pyramid of light, observed in the Mexican savannah.

1.6 MYSTERIOUS TWILIGHT

the heavens, connects both pyramids of light together. Even when it is stretched like a band across the sky, the position of this illumination distinguishes it from the Milky Way. It follows the zodiac and is therefore called the 'zodiac light', (From the Greek word *zodiakos*, meaning animal-circle.)

The zodiac light has long been regarded as one of the most mysterious phenomena of the night sky and even today there are many unsolved problems regarding its origin. It is a clear indication that space between the sun and the earth is not empty as was once believed, but that it is filled with a 'medium' that shimmers in the sunlight. Indeed, the counter-image and the connecting trail demonstrate that the delicate haze continues beyond the earth's orbit. Within the plane of the zodiac, it extends as far out as the distant planets and fills the whole of interplanetary space. Do we then see here with our eyes what we previously intimated, namely, the huge dimension of the sun filling space and enclosing the planets?

Indeed, not every astronomer is able to evaluate this perception meaningfully. Rather is it the case that a habitual earth-bound mode of thinking translates all our perceptions inevitably into the language of material conditions. Such conditions demand that interplanetary space be filled with a gas or with immense quantities of finely spread dust on which the light is reflected. In any case, all such implications require supplementary assumptions, partially contradictory, to back them up. Since the interplanetary gas can only be thought of as 'ionised', it follows that the 'dust particles' must be exceedingly small and 'very black'. Despite this, it remains inexplicable how this mass of dust could persist out there. According to calculations (Poynting-Robertson effect) it should have all been swallowed up by the planets, and particularly by the sun, long ago. Where is the source of replenishment? It is believed that the most likely origin is the disintegration of comets, the splinter products of colliding planetoids and the wearing out of meteorite material.

THE SUN

Figure 7. Early morning zodiacal light, photographed in Habakala on the Hawaiian Islands in January, 1967.

1.6 MYSTERIOUS TWILIGHT

Are we calmly going to accept the idea that what we are seeing are clouds of micro-meteorites and vagabond atoms, chasing round the planetary system? If we do, then we are closing the door on the possibility of finding any alternative understanding by making ourselves at home in the old earth physics. Why can we not linger a while in pure perception, long enough to allow it to lead us, perhaps, into uncharted lands of understanding?

Some scientists give us the feeling that they would like to crawl out of their skins, but that their academic thinking holds them back. Thus one writer (Stumpff 1957) states that the intensity of the zodiac light is in conformity with the intensity decrease of the sun's outer corona. But the ghostly apparition of 'interplanetary material' pops up again in the second half of his paper, though immediately thereafter he finds it necessary to add the restrictive comment that the spatial density of the zodiac light material is extremely slight. In the space between earth and sun there are 'only a few particles per cubic kilometre' and that one can see anything at all is due to the 'immense range of the zodiac light's cloud', which is described as flat, disclike and largely confined to the ecliptic (the zodiacal plane).

More significant is the train of thought of another author (Boschke 1970). He admits at the outset that 'science knows only very little about the zodiac light and the bridge of light to its counterpart'. However, he feels obliged to counterbalance this frankness with a radical proviso. One thing is certain, he says, 'The light which we see up there is reflected sunlight from meteorite material'. Then he gives a forcible description of how this material is not confined to a belt round the earth but (stretches not only to our neighbouring planets Mars and Venus but is distributed everywhere in the orbits of the planets, even enveloping the sun itself, where it is actually thickest!' The exclamation mark at the end of this quotation demonstrates how captivated the author is by his subject. He continues, 'It would be easy to assert . . . that the particles arose . . . today out of the limbs of the sun's

THE SUN

Figure 8. The feathery arms vibrate from behind the moon's darkend disk.

corona . . .' But then, he pulls up short with 'it is much more probable that,' and lands again in the meteorite dust and comet ash.

Why should we give priority to what is complicated and artificial, rather than to that which is obvious? An undaunted spirit is certainly needed to venture forth on the ocean of unaccustomed ideas and we have no intention of abandoning the province of logic and valid thinking. But if we want to move out into the ocean of the sun, we must be prepared to weigh anchor on earth's shore. We shall not have to avoid the solar wind, as we shall see in what follows.

To sum up then: We see in the mysterious charm of the zodiac

1.6 MYSTERIOUS TWILIGHT

light what our imagination had already intimated to us in earlier observations. The sun, extending far beyond what we usually take to be the centre of light and warmth, is also perceptible at night, for it is a cosmic entity embracing the whole. Our earth, like all the other planets, floats in the delicately lit medium of the vast sun body which, in its extended form, we justly call the 'corona'. Our earth moves on its orbit like a ship, lapped by the sun's waves. It is evident to us that the corona constitutes the essential nature of the sun and, as we have defined it, fills all space around the sun's orb, including the most distant planets.

As a matter of fact, we are astonished by the wisdom of nature that places sun and moon at just the right distances enabling us to see the corona, however seldom, when the moon exactly covers the sun's bright disc. The deceiving glare of the centre drops out in the same moment as the sun's true nature becomes visible. Is the whole harmony of the world directed towards addressing man – speaking to him in a symbolic language which would help him solve the closest riddle?

> So grasp without delay
> Holy open secret.
> > (Goethe, 'Epirrhema')

1.7 Mother and father sun*

Have we now ventured quite far into the regions of scientific debate? We must do some heart searching to ensure that each step in our thinking is fully human. On no account must we stray into abstract country where the spirit is left to idly wander. Every true act of knowledge enlists the heart.

Albert Einstein once insisted that every important scientific discovery be set down in a few key sentences on a sheet of paper. For him, the touchstone of lasting value is to be found in meaningful brevity. We would like to modify this criterion by asserting: All genuine knowledge acquired by means of head and heart together, can be so simply stated that even a child can understand it. For where heart speaks to heart, age and cultural differences play no part.

Let us enquire, then, whether the results we have arrived at above, could be explained to a child. If we find it difficult, we may turn to a poet for instruction. The mind of the artist is a sensitive indicator of what is going on below ground. As the seismograph makes visible the subliminal movements of the earth, so do the forms of a creative person tell us what desires to be brought to light out of the depths of the human soul. Long before an item of knowledge can be formulated scientifically, it is often

---

* The English poets, in the tradition of the Classic languages, see in the sun a 'father'. The Germanic languages see the 'mother' predominate. Both aspects are present in this omnipotent star.

1.7 MOTHER AND FATHER SUN

already present somewhere in poetic language, almost disregarded and not intended for professional instruction. The German poet, Christian Morgenstern, lets the sun sing a song:

> I am mother sun and carry
> the earth by night, the earth by day.
> I hold it tight and shine on it . . .

Is it not astonishing how aptly these childish words reflect the true circumstances? Where does the poet find this unerring clarity of spirit which, untainted by the fashionable thinking of his time, expresses what is natural and essential in simple sentences? Is it because he has made the last words of this poem 'Song of the Sun', his own life's maxim?

> Open your little heart, dear child,
> That we may be one light together!

Yes, here is a face-to-face knowledge that has no need of concepts. But who kindles this flame in us?

The company of children sometimes takes us so much out of ourselves that this inner flame leaps up. In such moments, we return to the transparent frankness of our forgotten childhood and give vent to a pristine imagination. Young people stir our emotions by their incessant demands and thus call forth the artist in us. When we talk to them our sober prosaic knowledge is transformed into ardent metaphor. Or we catch ourselves unexpectedly telling of hidden aspects of life we later discover with surprise to have been confirmed by specialists in some branch of science.

The ancient truth of the great mother star that carries and preserves our planet earth (and all her sister planets) in her radiant cloak can indeed stand the test of the large questioning eyes of children. We shall continue our adopted course accompanied by child, artist and prophet.

1.8 Sun hands

The peoples of early times were able to participate in an aspect of reality which today only lives in the wisdom of children and the inspiration of poets. The ancients lived in a harmonious world permeated by gods. They did not, as yet, suffer from the duality of belief and knowledge. Between spiritual and physical reality there was a bridge which they were accustomed to crossing, back and forth as a matter of course, and they expressed their experience of this intercourse between gods and men in vivid pictures and words.

Whoever saunters through the romantic Swabian university town of Tübingen, may come across the old unpretentious St Jacob's Church (or Spitalkirche) standing in an open grassy square in a quiet corner of the old town. In the medieval walls we find a carved stone on which there is a strangely impressive picture: *the sun with hands*. This emblem is high up on the south-east corner of the church, facing the direction whence the sun sends its rays during morning worship. Does it look upon its own reflection when its rays strike the wall — mother sun that carries the earth in its beaming hands by night and by day?

In this Romanesque church wall, what appears as an astonishing prophetic anticipation of modern knowledge may in fact be only the last echo of a mode of perception which was once universal, particularly among the pre-Christian peoples.

In earlier epochs, Germanic culture did not express its religious

1.8 SUN HANDS

Figure 9. The sun's hands in Tübingen. One of the three mysterious reliefs of the sun circle on the Romanesque St Jacob's Church.

or philosophic outlook pictorially. For this reason, much of what went on in the northern consciousness is hidden from us, and when a late fruit of this dim past is suddenly presented to us we are struck by its strangeness. However, closer study shows that it can be placed without difficulty in the larger context to which it belongs.

Other peoples have left us the content of their intuitive knowledge in a diversity of pictures and words. The Egyptians have left us an overwhelming wealth of such material. For them, both pictures and words were the mouthpiece of religion, and in those times that meant the whole of spiritual life. Here we find the sun's hands as a key element in the worship of Ra or Aton, spirits of the sun-god.

In a passionate hymn to the sun god, Akhenaton sings:

> Majestic is your radiance on the horizon of the heavens
> O Aton, creator, who lives from the beginning!
> When you rise in the east of the heavens
> Each land lies below you, shining in beauty.
> You are beautiful and great and sparkling and high over the earth!
> The lands and everything you have created
> *You embrace lovingly with your shining hands.*
> You are Ra that protects and guides them,
> That imprisons them in bonds of love.
> Though you be distant, yet your rays are on the earth,
> And your face is eternally turned towards her on your path.

And in all kinds of Egyptian sculpture we meet with a similar representation of the divine being of light emanating love. Rays reach down from the sun to the earth and its people below and each rays ends in a little hand. The great divine, celestial body embraces lovingly everything that it has called into life. It stretches over countries and oceans, caresses its creatures, and in the laying on of its hands, breathes into them its blessing. This is particularly well illustrated by the picture in which the rays of the sun, each ending in a little hand, distribute to all sides the mystery of life, represented by the holy sign, *ankh*, ☥, the key to the Nile. Thus does the radiant god protect and guide those souls devoted to him, as he does the whole earthly planet to which he continually turns his face on his path through the universe. The sun carries in its hands this human star that hurtles on its course without suffering harm and without falling out of the cosmos which coincides with the sunbeams of the central celestial body.

Could there be a more gratifying custodianship for the earth than to travel before the face of the omnipresent being of God,

1.8 SUN HANDS

Figure 10. The rays of the sun, as so many arms, shining upon Akhenaton and Nefertiti. Altarpiece from a shrine in Amarna.

illumined and warmed by his loving touch? Could there be a more inspiring and fitting description of the nature and activity of our mother star in whose bosom the vibrating life of our planet rests?

We see how the knowledge of ancient times coincides with the results of modern science. However, it will not have gone unnoticed that during our exposition, the description of the sun as a universal entity has become inseparably interwoven with expressions of a divine nature. Turning of the face, touching, caressing, holding and carrying, embracing — these are all terms of encounter used by a living, loving human spirit. In these expressions, the characteristics of the sun and of divinity are undifferentiated.

Figure 11. The sun as divinity within a wreath of rays as represented in a 1617 manuscript on the microcosmos and macrocosmos.

In fact, what the Apostle Paul, quoting Greek wisdom, said about God in his famous speech in Athens, can also be said of our guardian star: 'In him we live and move and have our being, (Acts 17:28). Whether we apply these words to the spiritual or to the physical universe, the outcome is the same. They constitute a basic triple chord and in this *one* truth, twofold reality can be heard.

We have not yet mentioned two other sun symbols on St Jacob's Church in Tübingen. They are certainly worthy of our consideration. In his book about Romanesque architecture Emil Bock (1979) compares the three sun circles. The two symbols without

1.8 SUN HANDS

hands are made up of a number of concentric circles and seem to allude to a movement in the play of forces of the sun's rays: contraction and expansation are apparently indicated in two different forms. Today, the three sculptured stones lie apart from each other in the wall. If, as is surmised, the reliefs once formed a single whole then the significance of this threefold symbol will be seen to have an astonishing depth and accuracy, the full meaning of which will become clear to us as we proceed.

1.9 Inside the sun

It is now but a short step to the realization that from the foregoing a profound conclusion can be drawn.

We have already said that the sun envelops the whole flock of planets in its radiant cloak, the corona, in its widest sense. We have made it clear that we float on our wonderful, living spaceship through the splendour of the sun's ocean, bathed and carried by waves of life-giving light. We have witnessed how observers of the corona during total eclipse are gripped by the joyous feeling that they are approaching the real nature of the sun when its mordant glare is covered up and the glorious rays of the corona fan out into space. Finally, we have noticed a similar mysterious presentiment of the nearness of the sun's true nature while observing the zodiac light at night.

Is not the time now ripe to assert that we live and move *inside the sun*? Indeed, these words seem to be the only valid conclusion we can draw from our thinking and they alone do justice to everything we have observed about the nature of the sun.

We are inside the sun. Notice how the usual conceptions are turned upside down when we conduct our research in accord with experience. At the same time, however, we must not attempt to lay down this difficult proposition as a formula to save ourselves the effort of thinking. As though a turn of words could make short work of our search for knowledge. Rather, we want to preserve such a proposition in our hearts, ready for action.

1.9 INSIDE THE SUN

One could of course object that the way we have spoken about the sun's hands contradicts the idea of being inside the sun. There is indeed a difficulty here which could be an incentive to widen our sphere of thought. We cannot doubt the reliability of the insight that has shown us the hands of the sun, nor the truth concerning the sun's interior. However, this interior has an exceptional characteristic. It cannot automatically be compared to the inside of the physical human body from which our sensations of inside and outside are derived. Normally, what goes on inside our bodies is unconscious and inaccessible. It is otherwise with spiritual interiors. And when we speak of the interior of the sun we are aware that we are, at the same time, speaking of the spiritual interior of a divine being. In the world of the spirit, we are conscious of inner events. It may be that the outside sets limits to the spirit and that, in the words of the Swabian philosopher, Oetinger, 'the physical body is the end of God's road'. Nothing is inaccessible to the persevering spirt that turns inward. When the hands of the sun reach inwards to encounter the body of their own light, then we see demonstrated a wonderful picture of superhuman, super-earthly forms of consciousness and life, towards an understanding of which we shall gradually move.

We could also try to understand the huge interior of the sun metaphorically, as a mother's womb in which its children are carried and out of which the world is born ever anew. But in this motherly womb there is nothing of the fertile darkness that permeates through the human organ of birth. It is a uterus of light in which the ripening life is nurtured by the caressing hands of the gods.

Sun, moon and earth — each contributes in its own way towards making mankind conscious of the all-surrounding, womblike nature of the sun, and thereby directing attention to the *true inwardness* of the supreme star that encloses the planetary system.

The sun by its blinding glare energetically repels attempts to perceive anything meaningful in its 'disc'. Sight is diverted to the environment of the light-filled earth where the observer encounters the reality of the sun's power.

The moon, through the singularity of its movement, conceals the disc's disturbing glare at rare moments and enables the rays of the corona to be seen fanning out into space beyond the earth's sphere.

The earth, however, by standing every night between the observer and the sun, makes it possible to see in the zodiac light the sun's shimmering interior, widely extended over the whole planetary system. As the spiritual content of sunlight was revealed to the old Celtic Druids in the shaded space of their stone circles, this mystery likewise is shown to the modern scientist in the shade of the whole planet when he 'looks at the sun at midnight'.

Step by step, we are being led into an intuitive perception of the sun's true nature and of the place of human-earthly existence in the cosmos: an understanding of our journey in the sun's inner space.

But now is the time to pose the question: what is there, then, beyond the disc where we formerly thought the fluid, fiery orb of the sun to be? If that place is so violent and continually recoils from our sight — should we not call it 'outer space' in contrast to the extended interior?

1.10 More about the corona

What is the relationship of the sun's orb to the corona's halo which surrounds it?

We learn from the astrophysicists some astonishing facts that have been brought to light by intelligent research. For example, they have attempted to ascertain the temperature at different places in the sun. The results are from every point of view surprising, perhaps even uncanny.

One reads that on the sun's surface, that is the brilliant disc we ought to be able to look at, there is a temperature of six thousand degrees Celsius. Similar research into the corona halo, on the other hand, comes up with a temperature of about one million degrees.*

These results fly in the face of all expectations, and have given the scientists much to think about. It presents them with a problem that is practically insoluble. The late astronomer Rudolf Kühn tells us (1962): 'The sun's corona is setting scientists numerous riddles . . . The most peculiar thing about the corona is its remarkably high temperature.' And now Kühn introduces an unforgettably absurd metaphor, also used by other astrophysicists. He continues: 'The question is: how is it that the saucepan is hotter than the flame that heats it?' He recounts a number of answers given in an attempt to answer this question. Formerly it was

---

* According to Dr Neupert the figures for certain places are as high as two million and in isolated places even five million degrees.

supposed that meteorites falling into the sun 'continually heat up the corona'. Today, the explanation is that 'soundwaves rising to the surface from the deeper layers of the sun's interior' would always heat up the corona. Such things are stated as though one could vouch for their truth, and yet, as Kühn guardedly but honestly admits, 'statements about the interior of the sun,' such as are made here, are 'pure hypotheses' and 'do not have anything like the same degree of certainty' as those we derive from direct observation.

In order to face up sensibly to this quaint metaphor of the saucepan, we must first of all ask what sort of meaning statements about temperature may have when applied beyond the sphere of the earth. This point must be dealt with before continuing our main topic, the nature of the sun's orb. (By custom referred to as the 'inside'.)

1.11 What do we mean by hot?

To put it in a nutshell: If we could convey a thermometer somewhere out into space and take a temperature reading, this would by no means imply that the reading corresponded to the degree of heat which a person might feel as the equivalent in bodily warmth. Why not? This brings us to the law of the specific stupidity of measuring instruments.

There was once a school of scientific philosophers who wished to apply this law, formulated more respectfully, to the efficacy of the human sense organs. They called it the 'law of specific sense energy' by which they meant that the eye cannot hear or smell, the ear cannot see or smell, and the nose cannot hear or see. All of which is irrefutable. Each sense perceives only its own allotted bit of the world. These shrewd philosophers hoped in this way to find an excuse for the limits of their capabilities. They maintained that, no matter how he tried, a person could not have any other conception of the world than that conditioned by the physical senses. They overlooked the fact that it is the very organization of the cognitive faculty itself that enables a perfect impression of the world to be gained out of many different though simultaneous sense perceptions supporting and complementing each other. Indeed, philosophers would render a valuable service if they could establish how man in his entirety is the only 'organ' that can perceive the world as it really is. 'But the present is the era of man. Today knowledge reigns supreme. For the first time since its beginning our planet, earth, sees and understands itself.' The geologist, Hans Cloos, begins his book (1954, 3f) with this enthusiastic declaration of man's greatness and mission which we could safely extend to the whole universe. By doing this we are most certain of staying on the road to reality.

However, the meaning behind the 'law of specific sense energy' is more appropriately applied to measuring instruments. Each one of these artificial tools of perception is monitored to register one particular impression and no other. Each gives us information only about one narrow aspect of the world. This cannot be compensated for by using many different instruments at the same time. They only offer many segments of the world placed side by side. But a sum of parts does not make a whole. Twelve slices of bread, however cleanly cut, do not make a loaf. Much less does a quantity of logs, be they ever so carefully joined, make a tree. And less again does a great number of instrumental readings give a picture of the world!

It is the intelligence which throws bridges from one particular to another, makes a word out of adjacent letters, picking out the meaning from between the letters (this is the meaning of *intellegere*) and often reading between the lines. A human being must bring his intelligence to bear to make a meaningful whole out of a mass of separate segments. In his excitement over ingenious measuring devices, man must not forget his own indispensable role as unifying agent. (It is the mind of man that integrates.) If, however, man takes his mission truly seriously, then the many diverse instruments can be very useful in sharpening and extending his sense perception in all directions.

What then does the law of the specific stupidity of measuring instruments tell us? It declares that a thermometer can only register degrees of heat no matter what happens to it. Even when it falls to the floor, or when it is hit with a hammer, the mercury moves and registers a 'temperature' which does not correspond to anything in reality. No matter what happens to it, this highly

1.11 WHAT DO WE MEAN BY HOT?

specialized thing can do nothing but react with a temperature reading, for this is its only form of expression. All other appliances behave in the same way, each according to its specialized nature.

In taking a reading from a measuring instrument, therefore, we must take all the accompanying circumstances into consideration. Readings by themselves are meaningless.

Strictly speaking, no matter how much apparatus we send out into the cosmos, we never get a true picture of conditions out there unless we integrate the information supplied by the instruments in terms of direct human perceptual observation. It is clear to everyone, however, that no human being can enter space under natural conditions. The astronauts have to be so tightly encapsuled in earthly conditions and their senses so well protected from cosmic influences that there can be no question of immediate experience. To rely purely on the information given by measuring instruments leads us nowhere.

To this, the expert can object that temperature records made in astrophysics do not rely on the simple readings of measuring devices, but are derived from an intelligent combination of and deduction from many different observations in the field of light and radio waves. It would divert us too far were we to go into the niceties of these technical points. However, it can be said in general that by the mere amassing of hypotheses, the immediacy of perception is made even more remote than by the 'good old measuring instruments'.

It is as though a man believed he could arrive at a distant and unknown part of the shore he lived on by crossing the water, travelling down a well-surfaced road on the other side and then recrossing. He was not aware, however, that he lived on an island and that the road in fact led further and further away from it. Recrossing the water he arrived not on his island but in a foreign country with no resemblance whatever to his home territory. But the traveller did not know this. He assumed he had come ashore in a land continuous with his own, though distant. Without further thought he describes what he sees, using his habitual concepts; and returning by the same roundabout route he describes his experiences to his fellow islanders, also in their narrow insular vocabulary. What a strange distorted picture of conditions in the distant land they will receive.

Astrophysics, which Rudolf Kühn soberly admits to be largely hypothetical, proceeds by many ways strikingly similar to those of our islander. What Kühn has to say concerning research into the sun's orb also applies to the investigation of the interplanetary region.

Following modern methods of thinking, physicists today are inclined to express the qualities of things in terms of motion. On the one hand, this is a justified attempt at uniformity, on the other however, a narrow specialization and a leaning towards conveniency. As a physicist once said (Lehrs 1951, 30) the attempt is made to make everything accord with 'the sight of a single colour-blind eye'. This shrunken, handicapped candidate for the cause of knowledge would at least notice when anything moved. He could also be easily replaced by a robot.

And so it actually has been contrived that warmth, that is temperature, is translated into concepts of motion. With this linguistic aid, manipulations can be made without difficulty, even in the cosmos. In the meantime, it is forgotten that on the 'other shore' one has travelled beyond the limits of the earthly facts of the home shore, and that translation into earthly language of what is met out there produces strange traveller's tales. 'Temperature' is spoken of where there is no possibility whatsoever of feeling any temperature owing to the changed spatial conditions. What is meant of course is 'motion,' but this word is unhesitatingly replaced by the word temperature as though what happened on yonder distant shore could, without further thought, be translated into terms of what happens on this side of the water.

Of course, clear-thinking scientists are fully aware of this problem. A questionable situation arises when popular hack

1.11 WHAT DO WE MEAN BY HOT?

writers, unscrupulously mingling word and picture, intend thereby to captivate their audience. Whoever tries to pin them down finds himself caught in a brutally materialistic world system, a universal skyscraper whose floors are undifferentiated because everything at the top is the same as at the bottom.

What, then, is the meaning of 'hot' in outer space? It means that, on the basis of a number of different observations, the conclusion has been reached that here and there more or less violent motions take place in the space-filled medium. The metaphor of the overheated saucepan is a dreary example of the depraved use of words whose presuppositions are ignored. For this reason it is comical, quite apart from the fact that it presents a false picture of the sun's spatial surroundings. We need not occupy ourselves with it any longer.

Our question is now very much to the point: what can we learn from the observation that in the region of the corona there is much stormier motion than on the surface of the radiant ball? And what conclusion can be drawn from this fact which would contribute to our understanding of the sun's kernel?

1.12 'That which is inside is also outside'

Between the numbers six thousand and one million there is a great difference. Purely as magnitudes, these values have little to do with temperature as we know it, yet their difference tells us something we shall try to understand.

It is like a clock that goes but is not set to the right time. Though we cannot tell the correct time by it, nevertheless, we can use it to measure the duration of an event. For if the mechanism works regularly, we only need to note the position of the hands at the beginning and end of the event and we can then calculate the duration. The actual time indicated by the clock is irrelevant for this purpose.

Let us take another example: Without accepting the ever varying figures which geologists give for the time lapse in earthly evolution, we can still learn a great deal about the course of the evolutionary epochs from a study of the sequence of these figures. Whether a particular fossil is forty thousand years old or a certain geological period ended forty million years ago, is quite beside the point. What is certain, is that the fossil marked 'forty thousand' is much younger than the find from the stratum marked 'forty million'.

We must learn from this to move away from the naïve esteem for 'absolute' numerical data and direct our attention to the relative values, that is the interrelationships of the time epochs to one

1.12 'THAT WHICH IS INSIDE IS ALSO OUTSIDE'

another. In this way, we are able to acquire another relationship to reality, not based on fixed, static assertions, but rather on a dynamic, mobile attitude, willing to adjust to the versatility of any event.

As it is with the hands of the clock that tell us of duration, and with the numbers that give us a criterion for judging the sequence of events in earthly evolution, so it is with the temperature measurements of the solar physicists; they give us insight into the relationship existing between forces in the sun. Since the degrees of heat really represent magnitudes of motion, we are, on the strength of these statements, advancing towards an understanding of the mobile inner ordering of a living space.

The fact that the flowing mobility of the outer radiance falls off towards the central orb, simply means that the sun is more powerful in the surrounding space than towards its centre. Just as one can say of a figure dancer that his very essence is in his limbs, and even extends into the wider circle of the awareness of the public, we must likewise imagine the essential being of the sun as existing most strongly in the extended encircling space, and weakest towards the location of the disc. Indeed, one is inclined to believe that towards the 'middle' this being disappears into nothingness.

From this new standpoint, then, we are again made aware of the sun's essential nature: the sun is more itself in the surrounding space than in the region of the orb.

Taking our earlier considerations seriously, let us now say it quite unambiguously. If the space in which our earth and the other planets move is the interior of the sun, and the activity, which we understand as a sign of presence, falls off towards the orb's surface, then we are literally compelled to accept that what lies on the other side of the sun's radiant skin is the exterior of the sun.

From the point of view of the sun, the earthly conception of 'inside' and 'outside' is reversed. The sun's interior is the extensive cosmic space in which we live. The spherical region behind the glaringly bright boundary surface is the inaccessible outer space. The sun puts a light barrier before the human eye, making it clear that its presence ends here: it is to be found everywhere, except where the fiery ring forbids entrance.

1.13 The hollow sun

The conclusions arrived at in the foregoing could be contested on the grounds that, whereas the temperature falls to 6 000° at the sun's outer skin, below this surface, in the ball of the sun, the temperature increases and is immensely greater than that of the corona. This unfortunately widely held view is one of those pure hypotheses of which Rudolf Kühn spoke so frankly. A more 'pure' hypothesis is hardly conceivable. It is the distinctive characteristic of the sun's skin that all penetration into conditions beyond it is barred, completely and for all time.

'We only know a skin deep layer of the sun's surface through optical observation and are unlikely ever to penetrate deeper with our telescopes.' This is the opinion of the well-known Freiburg solar physicist, Professor Karl Otto Kiepenheuer (1957), who is undoubtedly the greatest authority in the German-speaking world. What lies under this thin skinlike zone 'is for ever denied our sight.' *It is simply a hollow space as regards our sense perception.*

Using the means at our disposal, we can only make assumptions, conjectures and, of course, calculations. This is called 'extrapolation' which means that what has been proven here below applies equally there above. The solar physicist himself is able to express this more pointedly than we can: 'The picture we acquire of the sun's interior is a reflection of the current state of theoretical and nuclear physics and, like these sciences, is undergoing continual change. Solar physics and physics, therefore, are almost indiscernibly fused together . . .' The reality of the object of research is never even questioned. It is, in fact, a shuffling with words. Since the subject matter of research is 'invisible and totally inaccessible' one can calmly bury one's head in the sand, hide oneself in a physics laboratory and bluntly affirm that up there everything is the same as down here among the lenses, mirrors and electrodes. What the essential being of the sun really is, hardly troubles the solar physicist at all. He can proudly assert that he has managed to bring the celestial phenomena into step with the current state of terrestial physics.

But is 'extrapolation' possible on such an impassable barrier as the sun's outer membrane presents us with? Can we continue to use our customary concepts? Does this barrier not challenge us to pause in our research and to consider whether there is not perhaps another way of investigation more appropriate to the subject?

Furthermore, we ask: Is it because there is a 'hollow space in terrestrial experience' that we come up against an impassable barrier on the sun's skin? Is it because there is in that region *really a hollow space for all existence related to the earth*? And is not everything that has been thought up about the 'content' of this space a frustrating attempt to comprehend the incomprehensible?

When we hear of the 'fifteen million degrees in the nucleus,' and the inconceivable pressure and density, are we not reminded of the aboriginal brought to Paris, who, overcome with astonishment, called the Eiffel Tower a giant tree? He could only grasp what he saw by making an overstatement in the language of his accustomed environment. However, even the most gigantic and bombastic superlatives are inadequate to express what is fundamentally other. It is not therefore intended as a disparagement, but a rectifying acknowledgement, if in the records of solar research we see an impotent striving to understand and express verbally 'the wholly other.'

Considering the generations of effort put into research, it would

1.13 THE HOLLOW SUN

be arrogant to suggest that science ought to acquire new methods to deal with the otherness of this object and not at the same time be prepared to offer an alternative. In fact, unless one can point out at least part of this path, it should not be said that to attain knowledge of the sun, one must forsake the terrestial path and take the sun's path.

However, such directions along the sun's path can most certainly be given. Besides the perceptual and contemplative manner of observation we have exercised in the first chapters, we shall in what follows delineate a mathematical-geometrical method which will fulfil the requirements of an exact science. This cannot, however, be done in a few short sentences. Since we do not want to depart from sound common sense, it must be unfolded slowly, step by step. This will be done under the heading 'Opposing Worlds' in the second part of this book.

At this point we shall briefly recapitulate what we have discovered about the sun as the 'wholly other':

    the reversal of inside and outside;
    the picture of the parent star's mighty light-flooded body
        of rays, in which the planets move;
    the symbolic picture of the hollow space in the centre which
        defies any terrestial approach.

After we have asked one more important question about the mysteriously inaccessible exterior, we shall be able to make the transition to Christology.

1.14 Floating pearls

Can an empty space be visible? This question may well trouble many a reader attempting to think his way into the new conception of an inside-out sun. And quite rightly. For the brilliant membrane of the sun's orb appears so taut that one has the impression there must be a tangible substance behind it, filling out the space to bursting point. It would occur to a layman in a perplexed moment that a hollow sphere ought not to shine with such a solid, metallic glare; on the contrary, it ought to be transparent.

The answer, well known to opticians, may be discovered by anyone in a simple experiment. All that is needed is a bottle of soda water. How often have we not admired the silver pearls sparkling in the liquid! When the bottle is opened, they bubble forth incessantly. A little shaking encourages the phenomenon magically. Do we not sometimes imagine that we could take these glittering silver balls and hold them in our hands like so many pretty jewels?

But we know that the sparkling little spheres are bubbles of carbon dioxide, which make the soda water agreeable. Nevertheless, when we first see them, we always like to think of the fairy story of the floating silver pearls. What appears as a shining solid is an enclosure of less dense substance. Each bubble is a kind of hollow space in which a gaseous substance holds its own in a fluid. When we see the little bubbles in the water, we are looking through a dense material at a less dense one, and we realize that

1.14 FLOATING PEARLS

each hollow sphere is enclosed in a glittering silvery skin that prevents us looking into it and makes it appear like a metallic body. Or should we say that we are looking from the interior of the fluid on to the outside boundary surfaces?

At any rate, we learn from this the natural law that when we look at an object of low density through a denser medium, the plane separating them appears as an opaque reflecting mirror. Perhaps it would be a good thing to contemplate the soda water and its magical pearls several times, until we acquire a strong feeling for this law.

We shall then have no difficulty in imagining the hollow space behind the disc of the midday sun. We are looking from inside the cosmic body of light rays on to the sphere in the middle. Our vision is stayed by the bright reflecting membrane which prevents our entry into the mysterious hollow beyond. From this viewpoint the ball of the sun appears as the noble pearl, floating in the flowing light of the sun ocean. Is this the reason why its shining shell in its blinding brightness is so impenetrable? Not only because the boundary surface separates a dense element from a less dense element (as water from air) but also because of a transfer to a totally different spatial condition?

1.15 Cosmic whirlpool

Do you want to make an experiment and discover a natural law? If so, you can do a little experiment right now. You could do it at breakfast as you drink coffee, or in the evening when you brush your teeth, it's all the same. Start stirring in your cup or in your glass. Everyone has done this many times without thinking what a wonderful law of nature is being revealed to him.

The faster we stir the deeper is the funnel in the middle. We have seen this in the bath too. If we look into the glass from the side, we see a tube, like a trunk, winding down to the bottom. If there is anything floating around on the liquid, (say, cream on the coffee), then we notice quite clearly that the rotation at the edge of the vessel is slow and increases in velocity towards the centre. The nearer to the centre it is, the faster it whirls . . . It spins round with such speed in the middle that the liquid can no longer hold together under the tension created and is torn apart. The fabric of the substance cannot stand up to this strain and is drawn away

1.15 COSMIC WHIRLPOOL

from the vortex at the centre. A chasm opens up; this is the 'trunk' we saw from the side, boring its way down to the bottom. Naturally, this suction tube fills up with air which rushes in wherever a vacuum arises.

What would happen, however, if the whirlpool arose in a vacuum where there is no air? What is there to rush in when the thinnest of all substances itself is whirled apart and there is no other material that could rush in? Do we not have here a very graphic picture of mysterious antimatter, the youngest child of physics, still hidden by technical difficulties almost beyond comprehension?

Before drawing any further conclusions, let us seek confirmation of our own observations from the experts. In physics the vortex phenomenon is described as $rv = c$, (where r stands for the radius, the distance from the rotation centre, v for the rotating speed of the whirling substance, and c refers to a constant number). This is how a mathematician expresses the fact that the rotation speed increases towards the middle and in the centre exceeds all limits. When $r = 0$, the velocity is infinite. When the pressure (p) is investigated, it is found that when $r = 0$ then $p = \infty$. We are inevitably led to a negative pressure, that is a pressure less than a vacuum. (Not to be confused simply with pressure less than atmospheric). What is to be understood by this? To get some idea of what negative pressure is, let us refer to the theory of stress. Here, negative pressure is nothing other than tension. The same applies to a fluid. Generally, the fabric of the fluid ought to be torn apart prior to this.

What we have now learnt can be applied, almost as a matter of course to conditions in the solar system. Here, too, we want confirmation that our view is in accord with independent scientific research, and to this end we refer to the work of the gifted scientist Theodor Schwenk (1965, 44f). In his wide-ranging observation of the laws of flowing currents, he considers whirlpools in all their many natural relationships. He finds:

THE SUN

Figure 12. Oceanic whirlpool in the Pacific Ocean. This aerial photograph gives an impression of the great extent of the phenomenon.

On close examination we find that the vortex with its different speeds — slow outside, fast inside — is closely akin to the great movements of the planetary system. Apart from minor details, it follows Kepler's Second Law of planetary movement: a given planet circles round the sun as though in a vortex in as much as it moves fast when near the sun and slowly when further away. This law applies to the whole planetary system, from the planets nearest the sun to those furthest away. The vortex in its law of movement is thus a miniature image of the great planetary system. Its outer layers, like the planets furthest from the sun, move more slowly than its inner layers, which circle more quickly round the centre, like the planets nearer to the sun. The sun itself would correspond to the centre of the vortex.

1.15 COSMIC WHIRLPOOL

We would describe it otherwise in our newly acquired terminology: The whirlpool is a replica of the body of light rays embracing the whole planetary system, while the suction tube in the middle corresponds to the hollow space in the sun's centre.

Physical conditions in this centre surpass the loading capacity of material substance beyond all measure. The most attenuated ethereal medium cannot withstand such an immeasurably great suction, such abnormal stress. Physical substance bursts asunder under such impossible conditions. An absolute vacuum opens up. What takes the place of physical substance? Should we call it 'pure spiritual being' or 'antimatter' or 'ether' or 'antispace'?

In any case, we get a glimpse of a violent struggle on the boundary between this side and that side. The surface which appears so smooth and shining to the naked eye appears, through the telescope, as a bubbling, boiling cauldron and very grainy. These bubbles are called 'granules', but the expression is too neutral for the activity taking place over there. A mighty seething of currents, a conglomeration of furnaces, a clashing together of darkness and light, a collision of foaming tidal waves towering up one moment and thrusting down the next, a wild ebullience of thunder and lightning: all of which must be accompanied by the sound of a roaring hurricane in terms of terrestrial experience. This is what is shown us in the boundary zone where the opposing worlds of matter and antimatter rub against one another, collide and separate in eternal, incompatible turbulence.

We have now sketched a picture of the sun as it appears to our unbiased observation. From time to time, we gratefully resorted to the observations made by experts with accurate measuring instruments. We always tried, however, to preserve, as our base, the standpoint of immediate experience and a general approach.

It will hardly have escaped the observant reader that we were often strongly tempted to enter the sphere of 'Christology'. But for the sake of continuity, we have held back from crossing this barely noticeable borderline. There are perhaps, a number of thoughtful participants in this expedition who have made themselves independent of us and turned along their own path of discovery into this realm of spirit-soul experience. This experience arises wherever one succeeds in observing nature purely on the basis of a unified sense perception, namely as a complete human being. In the following chapters, therefore, it is only a question of filling in those omissions hitherto passed over; we shall tread those stretches of road previously overlooked and thus, step by step, reach our preliminary survey of the Christ-sun world.

First, however, we shall ask a question. It is of such importance that, before coming back to it in the third section, we want time to mull it over. The incisive question is: Does our conception of the great universal whirlpool offer us an explanation of the forces which hold the planetary system together? Can it be shown, as a result of our investigation, that Newton's hypothetical law of gravitation, which has increasingly become an article of faith, is a centuries old superstition? This law states that the central star keeps the planets orbiting round it by virtue of the force of gravitation. Can it be shown that it is a totally different power that controls the movement of the planets round the whirlpool centre?

We leave it an open question. May it cause consternation! If it gives rise to irritation, then there is a chance that in the painful desire for knowledge, a new idea will be born.

1.16 The sunbird descends

We have sought out the 'wholly other' in the cosmos and made its acquaintance. We have seen the winged sun whose nature is defined by its encircling environment as opposed to earthly things determined by reference to a middle point.

However, we know also that the two opposing spheres of activity do not merely lie side by side unaffected by one another. We can at least mention the desire of earth dwellers to grow out into the universe. Particularly today, many people are striving fervently towards a wider consciousness, with a determination to break down boundaries. It is a longing which we all feel, for confrontation with a being of cosmic dimensions, something all-embracing.

Is an overture also being made from the other side? Do we find anywhere in the world a sign that the sunlike being is approaching the earth?

We shall look at the great kaleidoscope of human history. There, at the axis of world history, we survey a scene which concerns us. It is preserved in a few succinct sentences in the four Gospels of the New Testament. It is the time when John the Baptist brought about a great change in consciousness among the people of Palestine by his fiery preaching and baptizing. The people streamed to him in multitudes. At an appropriate place in the Jordan valley, he set up a centre for his activity. There, he received people into baptism. Through baptism in the flowing waters, he absolved and eased the souls of a race who for centuries had suffered the strictures of an unyielding law while awaiting the Messiah. Through the cult of baptism in the river, he bestowed the inner power to begin a new life on those who were awakened by his teaching to the imminent arrival of the Messiah.

Then, someone came to be baptized whom John recognized as particularly well prepared. At first, he refused to baptize him, believing that it was rather himself that should be baptized by him who came. However, he perceived the historical necessity of the situation and carried out the ritual immersion. Immediately, he discerned the uniqueness of this baptismal event, which was the beginning of a transformation. Jesus of Nazareth did not remain under the water to the moment of asphyxiation as others did under the pastoral guidance of the Baptist. In his life's training, he had long since attained the threshold of death which was revealed to others below the water. For him, something else was to be inaugurated.

The Baptist noticed the spontaneous animation, he felt the devout integrity in the soul of the one being baptized, and sensed his majestic power to steer this act of initiation to its God-willed destination. Jesus had hardly dipped under the water, before he stood upright again. His ascent from the river called forth a response, something descended from the heights of space. The way had been paved for its descent along the channels of prayer. The priestly assistant now took on the role of witness to an event of universal significance. He saw the heavens open, and a radiant light descending on feathered rays enter Jesus and made its abode in him. A sound echoed in the surroundings and formed the words: *This is my son.* From the ends of the earth, the listening ears heard it. Today, the world's spirit has buried the sun's seed with productive power in the fields of earth.

The bystanders were gripped by the supernatural power thundering around the event. Some said there was a thunderstorm among the clouds. Others saw the river light up as though a

1.16 THE SUNBIRD DESCENDS

Figure 13. The Baptism in Jordan. Mosaic on the dome of the Baptistry of the Arians in Ravenna, sixth century.

mysterious fire glowed beneath the waves, sending its gleams far and wide. Yet others spoke of a snow-white dove which they saw flutter down to be united like a body of light with him who was baptized.

John realized that his mission had been fulfilled. He knew that he had been sent as a forerunner who had to baptize with water until such time as he was confronted with him who would initiate men's souls by touching them with the fire of the sun. He knew the infallible signs by which he could recognize the testifying soul in whom the cosmic fertilization took place. Now he had seen these signs and bore witness to the irreversible genuineness of the occurrence: *The sun bird has descended into the terrestrial world.* In Jesus he has found lasting materialization. He had seen it, and bore witness that this is the one who would steep terrestrial man in the radiant fullness of the sun's rays.

1.17 Pre-Christian vision of Christ

We can only understand the deeply felt influence of John the Baptist and the Gospels, if we visualize the hopes in which people lived before these events occurred.

It was against a background of general expectation throughout the earth that this unique event took place at the beginning of our era. The peoples of pre-Christian times all had visions of the coming of a divine light-bearer, given different names, bringing comfort and redemption from the suffering of existence.

Today, one can certainly picture those cultures of remote antiquity as being characterized by one basic underlying consciousness. No matter how magnificent some of them appear, nevertheless as time passed they became more and more enveloped in a grey veil of spiritual gloom. The intense love of splendour and the degenerate indulgence in luxury of many of the upper classes is best understood as an attempt to stifle the depressive consciousness of transience and inadequateness in their own epoch. The feeling that they were being subjected to the general decay of the world grew ever stronger. What was happening in the spiritual environment is best summed up in the words 'twilight of the gods.' Folk-souls were overcome by despair at the secularization of the once illuminating world of the gods, the alienation and darkening of an existence that had become rootless and homeless. This hopelessness urged the more ambitious to meaningless revolt and indulgence in bizarre and extravagant styles of living.

There were, however, circles in which a deeper consciousness was cultivated, and the course of historical evolution was followed. An ancient tradition was kept alive in schools of initiation and the hidden holy places of temples, where it was taught that this was a twilight period, a period of transition, and that when the lowest ebb had been reached, a redeemer would come who would infuse fresh life and introduce a new age. The simple people, who were tied to their drudgery and could not escape from spiritual poverty into worldly luxury, they too, harboured a healthy awareness of the changing times. They kept their integrity by holding on to religious teachings and practices where they were reassured that the night would end and the sun rise again. 'The people who walked in darkness have seen a great light' (Isa.9:2).

The Messianic expectation was particularly strong among the old Israelites. It penetrated the whole of life and governed its every detail. The faithful Jew never disobeyed the rule that upon wakening in the morning, he should look eastwards out of the window for the Messiah, who would come on the rays of the rising sun. For centuries, there were schools for prophets in Israel where the new generations of prophets were educated, who, by means of Messianic preaching and mighty deeds, would keep the eschatological expectation alive. It is both revealing and moving to see the struggles, both spiritual and worldly, between prophets and rulers, to hold on to the good life, to protect the folk-soul in its holy longing and fidelity to the redeemer prophecy.

However, our picture of the ancient world would be one-sided if we gave the impression that the Messianic expectation of the Jewish people was peculiar to them alone. Rather, we must see the suffering destiny and brilliant prophecy of Old Testament humanity as part of a worldwide historical movement, if we are to grasp the planetary dimensions of the pre-Christian Christ vision.

The progress of historical studies and archaeology is gradually bringing us a view which fills us again and again with wonder. At the beginning of this century, the theologian and authority on the

1.17 PRE-CHRISTIAN VISION OF CHRIST

Figure 14. Legananny Dolmen in Ireland, second to third century BC.

ancient Orient, Alfred Jeremias, wrote a book about extra-biblical Messianic expectations which is characteristic of the broad field of study compared with the earlier narrow Christian view. This is now an extensive field of research containing many branches of study. It marks out the framework of a modern understanding of Christianity in the necessary world-embracing light.

To see how extensive this study has become, we only have to place the ancient Egyptian image of Isis with the Horus child next to the Christian Madonna, or the virgin mother figure of the

THE SUN

Figure 15. The great relief on the Extern Stones in the Teutoburg Forest.

1.17 PRE-CHRISTIAN VISION OF CHRIST

druid initiation centre that used to exist at present-day Chartres (Richter 1965, Heyer 1956).

Let us consider the sun religion of the Celtic and Germanic tribes in Northern Europe. Today, we can still find many of their mysterious standing stones intact. Even though we have no detailed knowledge of the use these stone circles, menhirs, dolmens and alignments were put to, yet we have no difficulty in recognizing that all these initiation centres were in some way related to the sun. For Druid initiates, worship of the great celestial body was the means of determining their precise place in history. We learn from the Bride saga that the details of Jesus' birth in Bethlehem were known in Iro-Scottish regions long before the first reports began to filter through external channels. Through clairvoyant perception. the northern initiates perceived simultaneously all the sweeping events that happened in Palestine. They read in the sunlight conditions existing between earth and cosmos at the time. Dolmens and other sacred stone structures were their temples of knowledge where, in the secrecy of the shade, they consulted the subtle spiritual powers of light. They learned here the appropriate times. And it was in this 'solar research' that the historical moment was revealed to them, the moment when the spirit of light appeared bodily on earth. When, later, the Roman missionaries arrived in this region, they didn't understand the original sun Christianity of the Iro-Scots and set upon these supposedly 'heathen' elements with warlike ferocity. This brutal attack has considerably retarded the development of Christianity in Europe (Streit 1984).

We must not forget the Extern Stones in the Teutoburg Forest. In this central sanctuary, the Germanic tribes conducted a sun worship, traces of which still call forth our admiration, despite the deplorable ruination of the site. The surviving round window is indeed widely known. It is constructed so that it exactly faces sunrise at the summer solstice. Less generally known is the large pre-Christian sculpture of the god nailed to the tree which can be

Figure 16. Relief of the descent from the cross, Extern Stones.

1.17 PRE-CHRISTIAN VISION OF CHRIST

seen on the rocks at the same place. It was a custom going back into prehistory, among artists of ancient times, to make use of natural markings in the stone, such as cracks and splits, so that, by clever manipulation, they became part of the sculpture, or the artist was stimulated by the natural marks to see the sculpture. The Germanic god Odin, quoted in the Edda as having 'hung on the windy tree' appears as one of many mystical forerunners of the mystery of God on the cross on Golgotha. Charlemagne, known in history as the 'Saxon slaughterer', barbarically destroyed the initiation centres in the Teutoburg Forest in order to root out the religion of the Saxon natives. The raised chambers for observing the sun were destroyed together with other irreplaceable initiation symbols and constructions, in a campaign lasting several days. Surprisingly enough, the large sculpture of the figure on the cross survived unscathed.

Today, research is succeeding in probing the many-sided background to this missionary campaign. The large descent from the cross relief on another cliff in the same region, dated falsely to the twelfth century, must now be regarded as pre-Christian. Warriors were not the only adherents of Charlemagne. There were also representatives of a stream of initiates who, in a quiet way, attempted to ameliorate the atrocious effects of this 'mission with the sword'. It must have been these understanding men who encouraged the creation of the unusual sculpture of the body of Christ being taken from the cross. For it is placed next to the pre-Christian work as a sign that the ancient prophecy of Christ's coming had been spiritually perceived.

1.18 Expectation and fulfilment

Wherever Christianity understands its relationship to the general pre-Christian expectation of Christ, it brings the fulfilment of the great hopes for the future inherent in pre-Christian religions. It does not arrive as a foreign element in a hostile heathen world; rather the 'heathenism' is God's prepared ground, in which the seed of Christ can take root and grow. For, 'in its historical reality, the Gospel is not a new myth among old ones, but the consummated myth'. These are the words of the famous Protestant theologian, Adolf Harnack, and they carry particular weight insomuch as this flash of truth, coming out of his otherwise proven Christian blindness, appears as an irrepressible flame.

During the first two thousand years, the fundamental truth of expectation and fulfilment has been brutally violated many times, though even Jesus Christ himself, right at the beginning of his ministry, taught in the synagogue at Nazareth: 'Today this scripture has been fulfilled in your hearing' (Luke 4:21). All too often, missionary activity carried the cross as a symbol of fear and death into all parts of the world. Terrible scenes can be called to mind when contemplating the spread of Christianity. Plundering hordes, bent on massacre, march under the banner of the bitter, dark figure, destroying sacred modes of living and their monuments. We are only slowly waking up to the fact that the two planks crossed together embody only half the truth. The cross only becomes a symbol of blessing when it is linked with the circle of

1.18 EXPECTATION AND FULFILMENT

Figure 17. Irish cross with sun circle. Ahenny, County Tipperary, eighth century.

the sun. *Death's rune, encircled by the sun.* Here, the new reality becomes visual. Expectation and fulfilment. An additional source of power and a new degree of freedom is brought to the gravity and surface tension of our earth ordered by Christ, the light being who is born in its surroundings. In the ancient Asiatic, as in the Germanic, four-spoked wheel or sun cross, pre-Christian longing and deeply moving Christian truth merge into one another.

In his fresco *The School of Athens*, Raphael has given us an impressive picture of how the old and the new epochs meet and complement one another. According to Hermann Grimm (1889) the two chief figures are Dionysius the Areopagite, representing the Greek mysteries, and Paul, the Apostle bringing the new message of Christianity.

Dionysius points upwards. As an initiate, he knows the mysteries of the gods. Paul points down before him. He is able to proclaim that the whole fulness of divinity has now descended to terrestrial man. He bases his authority on the Greek statement already quoted about the divinity in whose all-embracing dimensions we live and move and have our being. The sunlike all-permeating divine being cannot be observed from outside and classified among other spirits. For this reason it remains, in the last resort, 'unknown'. As an example which stands for all time, Paul refers to the altar raised in Athens to the 'Unknown God'. In this way, he illustrates his preaching with incidents near at hand and guides his listeners from there into the new knowledge revealed to man: that the hidden god is now in a human body and wanders among us like a brother.

One of the most surprising prophecies of Christ's coming is to be found in the *Avesta* of the ancient Persian religion. It is expressed in a hymn with great clarity and beauty that the divine spirit of the sun will be transferred to a human being who will be a 'Redeemer' for all mankind, will heal the sick and overcome death:

1.18 EXPECTATION AND FULFILMENT

Figure 18. The School of Athens *by Raphael (Detail). Fresco in the Vatican.*

    The mighty one, bearing the royal promise,
    The sun-ether-aura, created by God,
    We worship in prayer,
    That will be transferred to the most victorious of redeemers
    And the others, his apostles,
    Who further the world,
    That enables them to overcome age and death,
    Decay and putrefaction,
    That helps them to eternal life,
    To thrive eternally,
    To free will,

> When the dead rise again,
> When the living conqueror of death comes,
> And through his will the world will be advanced further.

The translator, Hermann Beckh, knew the modern equivalent for what the Persian sun-worshippers meant by Ahura Mazda. It was the perception of the widely extended corona, that we recognize through our exact, modern methods of observation to be the true form of the sun. The ancient worshippers revered this sun-ether-aura. And as they raised their hearts to the great celestial body, they were, at the same time, aware of the eternal promise that one day, the sun being would sink down and incarnate as a human being so that the world might advance to a new future order.

Can one imagine a more illuminating preparation of the incarnation in the Holy Land than this ancient wisdom that perceived, through prayer, the mystery of the future arrival of the sun? For those who will learn the profundity of the Christian religion, is there a more enlightening link between the old world and the new reality? All religion teachers should urge their pupils to learn this hymn by heart. It is a wonderful method of stimulating growth of a feeling for the universal expression of God's will in the temporal sequence of expectation and fulfilment. The Gospels' teaching of Christ in all its detail, including the healing of the sick and raising from the dead, and thereby knowledge of the sacraments, is illuminated by the shining penetration in the power of the radiant sun. It raises religious life out of the narrowness of inner feelings into the breadth of the universe.

Let us now ask two further questions which arise from an accurate understanding of the Persian hymn text; the answers will emerge in the following chapters.

What is meant by the words 'the most victorious of the redeemers' which, in its use of the superlative, suggests one among many? And what is intended by the other plural which refers to 'the others, his apostles' who are also included as those in the

1.18 EXPECTATION AND FULFILMENT

sun's stream who would become human? Where must we look for the other 'redeemers'? Who is counted among the 'apostles'? Are we delivered from the spell of the once-and-for-all, incomprehensible being, inasmuch as this plurality leads us into a temporal succession? Does it refer to possible evolution, by showing us past and future stages of a 'sun way'?

1.19 The sun guardian as guest

In two ways we find traces of the great baptism by which the genius of the sun is wedded to the earth's body. However enlightened the prophetic religions were in expecting the transference of the sun-ether-aura to human existence, it would still be difficult for them to make the spiritual transition necessary in order to recognize, in the body and spirit of an earthly man, the divinity that was once perceived in the breadth of space. The ancient sun worshippers were too accustomed to the supra-terrestrial spiritual dimension of the light-being dwelling in the universe. They still looked outward to what lay beyond. It could not be otherwise. As the descent of the aura was gradually completed, their clairvoyant senses probed more and more into the emptiness. The closer the sun-being spiritually approached the earth, the less were the unchanged religious practices, directed towards the cosmic sun, able to find a response. 'The sun's throne is vacant.' It is in this plaintive cry that we hear the real change in the spiritual situation experienced towards the end of the pre-Christian epoch. 'Where has the enthroned one gone? How can we find the radiant guardian of the sun again?'

On the other hand we see the company of spiritually open souls who, perceiving his arrival on earth, flocked around him, whom they had so longed for. They found him in their midst, in conversation and round the communal table, shining like the sun. And

1.19 THE SUN GUARDIAN AS GUEST

they experienced how the life-restoring sun power beamed healing and transformation from his eyes and hands.

In his great novel, *The Brothers Karamazov*, Dostoyevsky (1982, 425f) describes the rise of this new spiritual presence as a wonderful event. In the fourth chapter of the seventh book, Alyosha enters the cloister after the revered elder Zossima has just died. He has been laid out, and the monks are keeping vigil in turns, reading continuously from the Gospel over the coffin. When Alyosha enters, a brother monk has just reached the second chapter of John. The subdued voice of the reader filters through the candle-lit room as the new arrival kneels down and fixes his gaze on the venerable face of the dead.

Alyosha's sense of sight and hearing is profoundly affected and he crosses the threshold into another spiritual dimension, where he sees himself as participating in the biblical scene described by the reader. There is the wedding table at Cana with the guests in festive array. And Jesus and his disciples are present. Suddenly, among the guests at table, he perceived the elder Zossima. 'So he too had been invited to the feast. . .'

> The elder raised Alyosha by the hand, and he rose from his knees.
> 'Let us make merry . . . let's drink new wine, the wine of a new gladness. See how many guests there are here? And there's the bride and also groom, and there's the ruler of the feast, tasting the new wine. . . . And *do you see our Sun, do you see him?*'
> 'I am afraid — I dare not look,' Alyosha whispered.
> 'Do not be afraid of him. He's terrible in his majesty, awful in his eminence, but infinitely merciful. He became like one of us from love and he makes merry with us, turns water into wine, so as not to cut short the gladness of the guests. He is expecting new guests, he is calling new ones increasingly and for ever and ever. There they are bringing the new wine. You see, they are bringing the vessels. . .'

THE SUN

Three thousand years before Christ, an Egyptian hymn to the sun praised the 'horizon Horus' as one who embraces the world:

> For you are the one that looks down upon the gods.
> No god looks down upon you.

The relationship of the sun divinity to the planetary gods was here perfectly described. Because they are incorporated in the light-body of the central star, they could not look at it from without. The planets are subordinate to the superior power and supervised by their cosmic spiritual overlord.

In the events of the Gospel, the sun is now transposed into human nearness. Its lofty radiance is softened. It shines out of the pure soul-being of Jesus as love and compassion. It moves through our lives without blinding or burning us.

Conrad Ferdinand Meyer has attempted to formulate this, the deepest mystery of our world, in the poem 'In the Sistine'. Here, Michelangelo engages in a dialogue with the divinity who had previously guided his hand in the powerful ceiling paintings of the Sistine Chapel. At first, Michelangelo sees him as a 'wild fire' in his macrocosmic creativity, permeating the universe, but then his impression changes:

> In my picture, it is you people
> Towards whom you mercifully float.

The act of creation as it is reflected in a Christian soul, is portrayed glowingly in two colours. In it, the sun-transformation appears as a primordial event, shining through all time, from nature's vehemence to the heart's light.

In the experience of soulful Russian piety, the sun event is ever present at the wedding table in Galilee. The true bridegroom continually invites new guests. He is there in our midst as the sun's guardian of joy and human kindness. In his presence, would water not turn to wine!

What the sun once did, when together with the earth's seasons

1.19 THE SUN GUARDIAN AS GUEST

it made the sap rise to glowing sweetness in the vine, is now done in the power of Christ's rays brought down to the human level. The guests at the wedding table experienced the moving of the spirit which earlier had been given them by the drink of the Dionysian god. Friedrich Hebbel says in a poem:

> Away with the wine! Who is not now inflamed
> Is not worth his kiss!

Today, we no longer need the stimulus of alcohol or other beguiling medications, which in ancient times helped souls to undertake the 'great journey' to macrocosmic divine experience. We find the sun-illuminating power of God in Christ, both in our midst and around us, now and always because it has itself undertaken to travel the earth. We meet it in the deeds and transformations of our fellow human beings, where it is not a question of more wisdom, but of more love.

1.20 The sun as procreator

Having presented the sun as background to the Christ phenomenon from different points of view, we can now return to a consideration of the Baptism described in the Gospels.

To begin with, we learn from *Matthew's* account that the immersion of Jesus of Nazareth differed essentially from all John's other baptizing in that Jesus needed only initial aid from John. Instantly, (for this word, the Greek text has *euthýs* which implies an inner energetic swiftness) he rose up and our expectation is aroused that now unique events are about to take place and must be accomplished by unusual means. The restraint, characteristic of the Gospel narrative even in the most dramatic episodes, requires of us special effort if we are to perceive the underlying goal.

In *Mark*, expressed in simple words, we find the same importance attached to the rising of the baptized and the descent of what comes from above. Whereas Matthew reports the event from a distance, as a miracle, this account is passionate and intense and at the same time factual. Here, it is not a matter of the heavens opening: no, the cosmic sheath is *torn apart*! What more precise terminology could have been used to express the great violence of the cleft between space and antispace which happens, as we saw earlier, where the fabric of matter breaks up with dramatic side effects? Through the fissure, which Mark opens up for us, the content of a wholly different world pours into the terrestrial sphere.

1.20 THE SUN AS PROCREATOR

When the dove descends from the heights, the relationship to the winged sun is immediately obvious to us. But Mark adds a nuance to the event. By means of a minor linguistic construction in the Greek, he unobtrusively directs the reader's attention away from the outer, physical aspect of the dove, and emphasizes the living movement in the beating wings. There is a difference between our seeing something descend 'like a dove,' and our feeling the movement of the wings in the dove's *flight*. In the first instance, we are concerned with the form, in the second with the moving course of the event. (In the Old Testament, too, where the dove hovers over the waters at the creation, the Hebrew expression puts the emphasis unambiguously on the hovering as movement. Alfred Jeremias (1911) is adamant that it is the movement to which our attention is drawn and that the dove, conceived as a static form, is a 'misunderstanding'.)

Luke adds a new element by mentioning a mysterious 'corporeality' which can hardly be translated. The physical aspects of the event are endorsed at a higher level once we have withdrawn from its crude exterior. This is because, according to Luke's account, the Baptism is revealed in its reality as procreative. Here, what penetrates earthly existence from spiritual realms brings into the creative forces of physical life a re-creation and a new creation. Corresponding to this is the prayerful soul of Jesus who, as a well-prepared fertile soil, is able to accept the sun's seed and thus form a bridge across which divine life finds its way into the depths of material being.

The creative generating power of the sun is expressed in the symbolism of both old and new religions throughout the world. C. G. Jung gives many examples of this. We read there, for example, that in the liturgy of Mithras worship the procreative spiritual breath issues from the sun, and this is expressed with metaphorical clarity as issuing from 'sun tubes' (Schütze 1960). It is clear how real this was to the ancients when one considers also the Indian Rig-Veda description of the sun as the 'single-footed one', or an Armenian prayer which describes the sun as resting its foot on the face of the person praying. There is also a carpet from the fifteenth century on which is pictured a pipe or tube stretching down to Mary from heaven and the Holy Spirit coming down in the form of a dove to fertilize the Mother of God. Undoubtedly it depicts the conception of the Jesus child. The relationship is obvious; at the Baptism there is a second, a heavenly conception, which will fill the mature soul of Jesus with the sun-descended Christ life.

Does this not bring to mind the 'universal whirlpool'? We only need extend the conception derived from hydrodynamics. From the hollow space in the middle of the planetary system a whirlpool tube now stretches like a feeler, and the sun's spiritual substance glides down and sinks into those chosen human beings whom it makes fruitful.

Thus everything is drawn together into one great unity. It is as though the sparse indications in the Gospels contained the whole wealth of mythical symbol and modern scientific conception. Their words are so chosen that they grow unaided in the receptive soul. The longer we allow them to live in us, the more they enrich us with knowledge and many-sided experience of the world.

To the facts stated by the other three Evangelists, *John* adds the important comment that the spirit of God 'remained' on the human being, Jesus. The spiritual contact is thus not of a fleeting, hovering nature, but allows firm hope of a lasting future influence. And that all this was and is the case is guaranteed in the fourth Gospel by a human being, a witness, John the Baptist, whose historical greatness and genuineness we have already met. The link with the social-judicial aspect of life is thereby established, and thus a genuine commitment is created between the cosmic scale of the event and its human proximity.

Whoever goes fully along with this and follows our line of thinking will come freshly to the knowledge that was self-evident in early Christian times, and only later fell into oblivion: that what

1.20 THE SUN AS PROCREATOR

was born and went forth from the Baptism, namely Christ, is the *Sol novus*, the new-born Sun-spirit itself. Common invocations of Christ were *O sol salutis* (sun of salvation). *Helios anatolēs* (sun of the rising) and the prophet Malachi, calling him the 'sun of righteousness' says that he will arise 'with healing in its wings' (Mal.4:2).

We know now that it is the sun-god, Christ, who dwelt in the man Jesus of Nazareth from the moment of the sun's procreation in the Jordan and that it is truly to him that all the characteristics must be attributed which in earlier times belonged to Sol or Helios or Aton.

1.21 The threefold trial

It would be remarkable if argument and conflict, even strife, had not arisen when 'the wholly other' entered the earth. Let us call to mind the violent friction that occurs continually on the sun's skin, where there is a decisive transition from the spatial region to the suction of antispace. What we perceive there as the external agitation of natural forces, we can expect to see in the sphere of human relations as storms in the soul, evoked by the presence of the spirit-sun.

In this light, let us consider the numerous disputes that Jesus had with the scribes and Pharisees. Think of the harsh challenges, appeals for amendment and calls for wakefulness that he had to sling in the faces of his narrow-minded, earth-cramped opponents. Again, consider such an incident as the so-called cleansing of the Temple, and finally, the furious attacks of the opposing side, aggravated into a war of extermination, leading finally to the Passion. Looked at from this point of view, all these Gospel narratives appear with a new lucidity.

In her delightful book, Dorothea Sölle (1968), the controversial modern theologian, investigates the way in which Jesus Christ faces the world. By relating in a lively manner everything that happened up to the present, she is driven to extol the inexhaustible moral imagination expressed in all the deeds of Jesus. He always does the unexpected, gains the victory over the attacks or malice of his enemies by turns of phrase and actions that one could never

1.21 THE THREEFOLD TRIAL

have anticipated. He escapes their perfidious traps, he presents himself as a living example of pure charity, unburdened by the hardened rules that chain man to the earth. And, with the wonderful sincerity of his words, he even heals the distempered bigotry of his opponents. For Dorothea Sölle, 'imagination is really a form of freedom'. She recognizes this through a profound study of Christ's life.

But where does this searching beam of freedom originate? If we bear in mind Christ's solar origin, then the greatness and beauty of his earth-rejuvenating imagination is not inexplicable. One who is from a totally different world is able to come as a 'liberator.'

It is through sun forces that Christ works, whether he breaks up hardened forms, or dispels the cancerous tumours of earthly existence by his radiance, or whether it be that he helps frustrated souls to find solace, and releases activity of thought. He does this by opening up 'wholly other' modes of perception and giving earth-born consciousness a cosmic breadth.

The admirable thing about the dramatic confrontation of sun and earth life is that the opposing forces do not destroy one another but, as though held by a placating will, gradually penetrate each other. This will to restrain, to sacrifice the sunlike being for the sake of an indissoluble union with terrestrial and human existence appears ever more active the further Christ's earthly life progresses. In this, we sense something of the profoundly inward and divine uniqueness of the sun being. It can display an immeasurable renunciation. Governed by immense devotion as it moves towards earthly man, it divests itself of its scorching, blinding, superior power and transforms any threatening gods, in a great sacrificial remelting process, into those earthly forces we could call compassion and love.

Right at the beginning of his earthly passage, in his encounter with satanic powers Jesus Christ underwent a threefold trial of this ability to change superhuman, violent forces into compatible human potency. The temptation arose at the point where the spirit, unaccustomed to embodied life, had to withstand the trials of earthly laws. These afflictions are of a threefold nature.

The experience of hunger is novel and unknown to a spiritual being. To begin with, the son of heaven had to learn that metabolism was essential if the earthly body were to endure. Here, the tempter thought he would be susceptible: 'If you are the Son of God, command these stones to become loaves of bread.' But Jesus turned aside the request that he should appear as a popular minister of nutrition. 'Man shall not live by bread alone . . .' (Matt.4:3f). Indeed, it is even true that bodily satiation leads to sluggishness and apathy in the soul. Christ leads his disciples along another way. He leads them, through the intermediate stage of the 'feeding of the five thousand', to the inauguration of the communion on Maundy Thursday, and shows them thereby a path of development in which nourishment becomes increasingly the bearer and mediator of spirit forces. In this way, he plants the first sun germ in earth's life.

A spiritual being is also unaccustomed to the laws of gravity. That all material should be subject to weight and drawn towards the earth's centre is foreign to the experience of the sun being. The tempter sets Jesus on the highest pinnacle of the temple and says to him: 'If you are the Son of God, throw yourself down . . .' (Matt.4:6). But Jesus is not tempted to suspend the laws of nature by acts of magic. He knows that even the greatest working of wonders does not really help people, but only leaves them crippled and discouraged in their souls. Christ transforms the temptation and inserts a second sun germ in the being of the earth. He advances along the way that overcomes death, that raises others from the dead, until he reaches his own death on Golgotha. Thus he sets humanity the example of how, in death, the spiritual being of the 'I' can be saved from the decay of the body. Where the earth reclaims the body's substances it will be essential that each individual acquires the power of spiritual resurrection.

For the first time, a spiritual being must also become acquainted

1.21 THE THREEFOLD TRIAL

with the conditions of possession. Human beings find difficulty in understanding that 'perceiving' and 'having' can not always overlap. Observe how painfully the little child learns this hard earthly principle. Even at a mature age, we often don't find it easy to distinguish between what we desire and what is actually possible. The tempter's power gathers itself for a final onset. On the summit of the mountain, the initiate is offered a gorgeous symphony of sense stimuli, the dazzling beauty of which threatens to blot out the borderline between illusion and reality. 'All these I will give you, if you will fall down and worship me.' (Matt.4:9). But Jesus does not succumb to the temptation of at once acquiring for himself everything that his eyes survey. Such violent abruptness would not be in harmony with the self-determination intended for the future of man. Christ begins to build up his kingdom so that it grows slowly, and eventually embraces the whole world. Only in this way could he preserve the freedom of humanity. The kingdom of Christ can belong to no-one through external compulsion, or by receiving a stamped membership card. It belongs to each person only in as much as he allows his heart to be warmed by the sun's power streaming to him through Christ from the hearts of other human beings.

Contemplation of Christ's deeds calls forth our highest reverence for the fathomless greatness and virtue dwelling in this, the divine majesty of God, the manner in which he makes decisive victories in three fundamental spheres, not by violent means, but through the subduing and transforming spirit of the sun.

1.22 Cosmic consciousness

In principle, one single sentence in the Gospels suffices to make us aware of the full meaning and content of the Christ-sun-mystery, namely, the words in which the God who became man expresses his own nature — 'I am the light of the world'. It is plain that this utterance can only be true and genuine in one particular instance, that is, when it issues from the lips of the sun itself. Here, it is immediately obvious that what we experience on a grand scale externally as the reality of the cosmos is identical with what we experience in human intimacy as Christ, the light of the soul. Let us therefore examine the context of this sentence (John 8) for its broader meaning.

In the first place, the light-Logos appears in its transparent structure as a regulating prism whose radiant beams bring clarity into whatever lies in their path. 'I am the light.' Here, the cosmic consciousness, in which the 'I' of Christ identifies with the light being of the cosmos, comes towards us. But 'cosmic consciousness' in terms of the sun, is not only characterized by universal expansiveness, by superabundance; it includes, at the same time, a strong, warm predisposition towards the earthly world. 'I am the light *of the world*.' (The meaning of the Greek word *kósmos* contains orientation towards a goal.) We are led back to the self-evident primordial wisdom of the ancient world that never doubted sun and earth once formed a unity that gradually broke up as the planet earth, growing increasingly harder, fell away and

1.22 COSMIC CONSCIOUSNESS

separated itself from the living, flowing stream of life. This is what was originally meant by the 'Fall', the fall into separateness.

Now, however, through the descent of the sun-being, a new connection is established, sun and earth begin to draw together again. This is contained in the reflective double meaning, 'I am — the light — of the earth-world.'

Why is there such general gladness at the beginning of a sunny day, when the radiant daytime star, in a cloudless sky, meets the earthly landscape face to face? And why does it put us in such an ill-humour, depress us and enfeeble our creativity when a blanket of fog perpetually hides our view of the sun?

For the very good reason that sun and earth belong together, because originally, they were a unity that only later broke apart. Therefore we experience the pain of separation ever anew when an obstruction intrudes between these kindred stars. Therefore we are filled with longing for the restoration of that primordial unity, and with joy when there is bright interplay between them.

We know of a similar radiant joy in a quite different sphere of experience. Our souls are moved when we see two people discovering each other and joining in lifelong matrimony. And we are depressed when we see two people separate, and break up communal life under dark influences. Why do we feel this joy, when destinies are joined, this pain, when the ties of loyalty are broken? Because every human union reflects, in miniature, the primordial cosmic unity of the stars.

In earlier times, therefore, the punishment meted out to those who committed adultery was gruesome, for it encroached on a universal process, and human behaviour could either augment the separation or gradually bridge the gaping cleft in the cosmos. The Mosaic Law commanded that adulterers be stoned. It is as though those who did not have sufficient endurance and stability in their inner lives, no firm principles or harsh self-discipline, needed to be compelled from without by hard stones, meted out as a terrible medicine.

A different cure for the same human weakness is described in the eighth chapter of John. The scribes and Pharisees bring an adulteress before Jesus and ask him what should be done with her. Jesus' reaction is quite unexpected. He bends down and writes on the ground with his finger. As they question him, he already knows how to prevent them stoning the woman. And again he bends down and continues writing on the ground.

What could he have written on the earth? A particular word or sentence has not been preserved. We can only say that he wrote his own self into the earth's book. A writer always imprints the writing material with something peculiar to himself. (A study of handwriting can reveal something of a writer's character.)

Did Christ perhaps write 'I am the light of the world'? At any rate, wherever he touched the earth, he left the imprint of his divine sun-nature. And thereby a new unity of earth and sun was established, a cosmic marriage union pledged. Through this deed, the two worlds that were moving ever further apart were given a perceptible content to bring about a future drawing together. This is the cure that Christ offered for the break-up of earthly unions. He did not acquiesce in individual failings but went to the root of the evil in lucid divine greatness. In his presence and deed, the rift in the world is already bridged, and there lies the power that overcomes the demon separating people.

Through this event, a foundation stone is laid for all time. Whatever the individual's destiny, be it in distress, be it in the privation of all loving relationships, or in pining longing, one fundamental point has been won: the god who can say of himself, 'I am the light of the world', has been on earth and stamped his sun-seal on our planet. Thereby every common destiny is lifted up anew into the light of the stars.

This unique sun sentence from John's Gospel shows, in the profound meaning of its context, that it is not only significant for knowledge of Christ but is also incalculably active in everyday life.

1.23 The sun path

And the others, his Apostles, what about them? The life of the sun-spirit in human proximity is itself a promise: 'he who believes in me will also do the works that I do' (John 14:12). This is the message and the deep meaning of the divine sacrifice through which the spirit of the cosmos so enters the image of man that a *sun path* will be paved that is open to all.

There was never any doubt in olden times that a path into the macrocosm existed. But the inner sequence of stages to reach it was undertaken only by an elect few. A careful choice was made in the temples and schools of initiation to ensure that only the best were prepared for the highest possible degree of initiation. A title signifying the completion of his initiation was conferred upon the individual, according to the progress he made towards the divine status. Such titles were 'Messenger' (*angeloi*); or 'Raven', 'Guardian of the Mysteries' (*occulti*); 'Spiritual Fighter' or 'Warrior'. There were 'Sun Guardians' and 'Sun Runners' (*heliodromos*); there was the 'Saviour' (*sōter*) and the 'Anointed' (*Messias* or *Christos*).

All these wonderful distinctions attained by human beings were, at the same time, prophetic anticipations of what was to come. And an occult melancholy overshadowed the whole of this pre-Christian initiation system because one felt that the higher development of these outstanding souls was only attainable at the cost of a vast number of 'underprivileged' human brothers remaining behind. The sighing, longing expectation of the best, was, therefore, always accompanied by the conviction that one day the 'most victorious of the saviours,' the *verus Christus* would come and establish a breakthrough for the path to the sun that would no longer be exclusive, but could be trodden by all.

The guidance that Christ gave humanity was accepted by the early Christians with joyful, courageous enthusiasm and it was common, therefore, in the early period, to address those who fully belonged to the community as 'saints', and thereby place them close to the angels. Both angels and saints have the sunlike shimmer round their heads to express their spiritual nobility and show that here a being, illuminating the soul, is shining. (It is similar with the king's crown, a symbol of an augmented consciousness. The bearer is not only responsible for his own destiny but for the weal and woe of a people.)

The members of the first Christian communities experienced themselves as committed ambassadors, as 'missionaries,' as 'apostles' whose mission it was to bring healing and a new world to their fellow human beings. As heralds of the good news, they were 'angels', that is messengers; as disciples of their sun master, they were 'sun runners'; as healers of the sick world, they were 'holy' or 'saviours'. Indeed, they could not grasp the immensity and novelty of what they experienced otherwise than by calling each sincere believer a 'Christian', that is a 'Christ'.

What is the sun path?

If to begin with we wish to give a simple description of this, we could do it in the words of the sun itself as expressed, you may remember, in Christian Morgenstern's little verse.

> Open your heart like a little cup
> For I will also shine in there

1.23 THE SUN PATH

> Open your little heart dear child
> That we may be *one* light together!

If only we human beings would feel more strongly the impoverishment we suffer when we leave the realm of childhood! We would then not only guard against compelling children too early and too quickly into our dry adult world with its pretences of maturity, but remembering our loss of a richer, more lively world, we would also exert effort to discover the path to youth. This path is, at the same time, the Christ path: 'Unless you turn and become like children, you will never enter the kingdom of heaven' (Matt. 18:3). The child's verse on 'Mother Sun' is in full agreement with this fundamental teaching of Christ and serves, therefore, as a valid rule of discipleship for the seeker who would take the sun path. However, we want to mark the way with more clarity and depth.

Whoever knows that he lives *in* the sun need not regard the goal of his spiritual journey as a distant foreign land, rather it is a question of opening those organs which allow him to perceive what already surrounds him — this is the meaning of 'opening the little cup.'

Jacques Lusseyran whom we have already quoted as an initiate of light, describes in fascinating modern and actual terms, the development of new light-sensing organs for perceiving the reality in which we live. He speaks for himself, but he emphasizes that the way he has found can also be taken by anyone.

Physical blindness led Lusseyran, at an early age, to seek some other way of experiencing light. His determined search and practice was not in vain. It was only after he had given up the earlier mode of perception, which was directed outwardly to the surface of things, that he learnt a totally new kind of perception, somewhat resembling an inner listening. Once again light became present for him (1963, 10f): '[I] bathed in it as an element which blindness had suddenly brought much closer'. Indeed, he was brought into a closer and more intimate relationship with the light than ever before.

> . . . the opposite of light was never present. Sighted people always talk about the night of blindness, and that seems to them quite natural. But there is no such night, for at every waking hour and even in my dreams I lived in a stream of light.
>
> Without my eyes light was much more stable than it had been with them.

What was the reality which the 'clairvoyant' blind man grasped with his new organs? Not dreamlike, by any means. Jacques Lusseyran's comrades knew very well that he possessed a most exact sense of orientation. If the boys lost their way in the Parisian labyrinth, they would ask their blind friend the way home. It also happened later in life, that he would often tell the taxi-driver which turning to take. When he went along an avenue, he could demonstrate his 'keen-sightedness'. He could unerringly point out each individual tree along the side of the road. In his youth he went for long walks in the mountains with a friend. On these occasions, the friend who could see rested a hand on his shoulder and by means of pre-arranged signs led him over stones, hollows and other obstacles without having to disturb the silence with unnecessary words. In this way the one with sight led the one who was blind over the rough ground. But it was the blind man who drew attention to the great beauty of the landscape; how a group of cliffs reared up yonder, how upright a wooded copse grew here, how far below a village lay in the valley, or how gracefully the river meandered.

Do we not see here a human being who has burst through his body's bonds; how he crosses them in radiant unity with his surroundings? Ought we not straight away bestow the golden shimmer on the head of one who is taking the road to sunlike omnipresence, in token of the expansion of his being?

> Still, there were times when the light faded, almost to

1.23 THE SUN PATH

the point of disappearing. It happened every time I was afraid . . .

Anger and impatience had the same effect, throwing everything into confusion. The minute before I knew just where everything in the room was, but if I got angry, things got angrier than I. They went and hid in the most unlikely corners, mixed themselves up, turned turtle . . . This mechanism worked so well that I became cautious . . .

I could no longer afford to be jealous or unfriendly, because, as soon as I was, a bandage came down over my eyes, and I was bound hand and foot and cast aside. All at once a black hole opened, and I was helpless inside it. But when I was happy and serene, approached people with confidence and thought well of them, I was rewarded with light . . .

Armed with such a tool, why should I need a moral code? For me this tool took the place of red and green lights. I always knew where the road was open and where it was closed. I had only to look at the bright signal which taught me how to live. (Lusseyran 1963, 12f).

On Christ's sun path, the criteria of judgment regarding a person's actions become increasingly his own responsibility. 'You have heard that it was said to the men of old, "You shall not . . ." But I say to you . . .' (Matt.5:21f). And this teaching of inner sovereign freedom in responsibility for others reaches its climax in the sentence: ' "You shall love your neighbour as yourself." . . . love is the fulfilling of the law' (Rom.13:9f).

What power is it, then, that ultimately brings about this increase in being? ' . . . an end of living in front of things and a beginning of living with them. Never mind if the words sound shocking, for this is love.'

And how do we exercise this transforming and eye-opening power in our lives? 'And this continuing miracle of healing I heard expressed fully in the Lord's Prayer I repeated at night before going to sleep' (Lusseyran).

The practice of prayer appears in a new light during the course of the sun path. Whoever lives within the great, world-embracing shining of the light, has no need to call impotently upon a distant God. It is again only a question of opening doors or windows so that the omnipresent, flowing light can stream in. It is a question of putting obstacles aside and removing barriers. Prayer opens doors, makes dull windows transparent, forms more sensitive, far-reaching sense organs. 'For your Father knows what you need before you ask him. Pray then like this: Our Father . . .' (Matt.6:8f). It is a question of swinging ourselves into the great cosmic movement of the colourful, life-giving source of the light which bears all existence along.

Where does the sun path lead?

Despite the scantiness of the accounts in the Gospels, the goal is presented to us with unmistakable clarity at that point in Christ's life where the sun nature is openly revealed in him, namely, in the Transfiguration on the mountain. Here we see Jesus Christ at the turning point of his earthly mission. 'His face shone like the sun, and his garments became white as light' (Matt.17:2). He had reached the highest point attainable by earthly forms of life: Physical matter became filled with light, and shone.

Without the power of Christ, this climax of perfection would be at the same time a terminus for a human being; it would be connected with death. For the son of God, it is now only the beginning of his suprahuman mission.

The initiates of pre-Christian times presented humanity of old with an example of the spirituality which would lead to redemption from the earth. We see this most clearly in Gautama Buddha who taught how the 'craving for existence' was to be overcome and his followers thus saved from being enmeshed in existence.

Christ's path leads beyond personal fulfilment and redemption.

1.23 THE SUN PATH

Figure 19. The Transfiguration, *by Raphael, Vatican.*

His direction points towards the earth. He is the path of transformation. Raphael has expressed this mystery wonderfully in his picture *The Transfiguration*. Christ hovers. He is received into their realm by Moses and Elijah, representatives of the great leaders of mankind. One might think he wanted to leave the earth and join them. But his will is contrary to this. He does not float away. Nor will he, as Peter thinks, linger in the heights where he may be worshipped from afar as a strange, distant god of miracles.

He descended into the valley. And his heightened power, attained through renunciation, is revealed in his healing of the lunatic boy. The disciples had not been able to do this. Christ opposed the power of the moon, which influenced the boy, with the power of the sun that was his nature. Thus begins the great work of earth transformation which, with death and resurrection, puts its mark upon our planet, and should be continued by those human beings who accept it.

Christianity is the faith that loves the earth.

2
Contrary Worlds

Go to the devil with your duke, and leave me in peace to contemplate the Divinity, lit up by his own sun.
                              Joseph von Auffenberg

2.1 The opposing players

It often happens in life that a contrasting example serves us best when we are seeking clarity. Contrasts limit one another and emphasize each other's specific qualities. This is the case with the sun and the earth.

These celestial bodies are of such a contradictory nature that for every assertion we make about the one, the opposite will be true of the other. In making such assertions, a specifically close relationship between the two is revealed; the partners are bound to one another by being diametrically opposed in all respects. Such a relationship is called a 'polarity'. As the poles of the earth are spatially opposed, the 'polar' ends in an electric field likewise present forces in a relationship of opposition; this applies, for example, to the poles of a magnet. Here, we will study it in conjunction with the polarity 'sun-earth' in cosmic space.

Let us first of all describe the structure of the earth. No matter where we are, we can always say that we have the huge body of the planet under our feet. If we happened to be in a deep valley with the mountains rising steeply to either side, we would still know that beyond their summits the land falls back again. And even if there are higher mountains, the greatest mountain range is but a small uneveness compared to the ball of the earth, so we are never deceived. Basically, wherever we are, we are 'on top' of the earth, on a knoll, as it were, and the globe rounds off below. All the weight and heaviness of physical matter is contained in

THE SUN

Figure 20. Photograph of the earth in space, taken from Apollo 11. *The thin film below the line, between the arrows, is the atmosphere.*

the spherical skin of the earth's surface. There are traces of earth-like substances outside, it is true, but they soon pass away. One is astonished, when looking at a picture or model, to see how thin the earth's atmospheric skin is.

As we have seen, quite the contrary is the case with the sun. Its nature is such that it can no more be confined within a boundary than the earth can stretch out beyond its own boundary. Nothing at all corresponding to the earth's surface is to be found on the sun. The radiant skin of the sun is something quite

2.1 THE OPPOSING PLAYERS

different. According to our observations, we must assert to begin with that the entire sun extends to an incalculable distance, is so limitlessly distributed round about that we are unable to give it any definite spatial dimensions. In the case of the earth, a spatial dimension can be derived from the circumference of the globe. But what can be said of the sun? Admittedly, we could say that it is as big as the whole planetary system. But how big is the system? Earlier, Saturn was thought to be the most distant of the planets. In more recent times, Uranus and Neptune have been discovered and recognized as members of the family of planets. Today, Pluto is acknowledged to be the most distant brother among the celestial wanderers. Nevertheless, we know of comets that reach out much further into space on their extensive orbits and which we must also reckon among the community of sun related bodies. But having said that, what have we established? Even though we knew the most distant comets and could give exact figures for the extent of their orbits, would it be certain that the influence of the solar system did not stretch beyond these outermost orbits? Where is the dividing line to be drawn for the sun's radiance that floods the cosmos?

We notice that we and our conceptions are driven out of our conventional spatial feeling when we link up with the peculiarity of the sun's structure. Thus the most simple spatial measurement, distance from the sun, is really not possible. Conceptually, we can indeed set up a measuring rod on the earth. There is a clearly defined foot-rest for it. But to which point are we going to measure? The sun does not offer us any such firm foot-rest for the measuring rod. It is an insoluble problem. Somewhat as though we were asked to measure the distance between Berlin and Europe. How shall one measure, when the one is contained in the other? Between a planet and its all-encompassing sun's womb there is no distance that could be expressed in measurements of length.

One could use the subterfuge of measuring the distance from

the earth-ball to the 'sun-ball'. But did we not find earlier that the sphere in the centre of the sun star is not 'the sun' at all, and that the radiant skin which blinds our eyes is only the boundary of the inessential, of the outside? In this way, we could indeed arrive at a measurement of distance, but what these calculations have to do with the sun's reality remains doubtful.

The 'opposing players' are shown in the following picture: The sun, immeasurably extended over everything. Within the whirlpool movement of the radiant body, the firm material body of earth orbits together with the other planets. Sun in circumference, earth in itself. The massive earthly interior shuts out the light just as the hollow beyond the whirlpool's centre withdraws from the intrusion of gravity-bound intelligence.

Man lives *on* the earth *in* the sun. He is placed in the midst of a field of dual forces between cosmic opposing players. How does he fit into this situation?

2.2 Man as mediator

Man's mission, to link the realms of earth and sun, is expressed in his upright posture. In this respect, the characteristic of human beings to act as a bridge is unique, and quite different from the vertical growth of plants. The plant stems rise as a definite line into space like the human vertebrae. Flower and root appear as polar organs. However, inasmuch as the root goes down into the centre of gravity within the earth and the florescence opens upwards into spatial light, warmth and movement, the duality is threatened; do we not sometimes feel that a fragrant flower is about to flutter away like a butterfly? In man, the encounter of what is above and what is below blends together and is augmented by mutual penetration. Let us take a closer look at the polarity between head and limbs.

The human body has a piece of the hard mineral earth in the skull. The head contains the hardest bones in the skeleton and its shape is comparable to the earth's globe. It is that part of the body which, by nature, is characterized by petrification, impenetrability and self-sufficiency.

The limbs are quite different. While one can say that the head contains everything hidden in a capsule, the limbs appear only as appendages to something which is by no means complete in itself. Because of the way they are formed, arms and legs point in different directions, they point away from the centre and express a liveliness directed towards the environment. The way in which the skeleton is attenuated the further it stretches out from the trunk, is a visible demonstration of its persistent striving to exert an influence on the wide world. First, there is one bone in the upper arm and the thigh, then two below the elbow and in the calf of the leg, then there are many little bones in the palm of the hand and in the foot, raying out into the five fingers and toes. This branching out of the bodily form is an expression of its desire to dissolve into its surroundings, to merge into them so that it is no longer differentiated, but has become one with its surroundings. The extremities of the limbs, fanning out, bear unmistakably their relationship to the sun's rays filling space. (Should one not also say that in a contrary fashion, the posture of the limbs shines from the infinite distance into the trunk where they have become more and more closed in and solid?)

The polar characteristics are clearly defined. What is surprising is that in the human body they are so remarkably changed about. The head is raised above and the limbs are turned towards the ground.

It is not just as a matter of course that man goes through life with his 'head high'. When he is born he enters the world the other way round. The child is carried head down in the womb and falls like this into the gravity of the earth. In fact, ought one not regard this position as natural, considering that the head is heaviest and most similar to the earth? But a counter force begins to stir forthwith in the child's life. Newborn babies try to lift their heads already in the first days. The train of events then initiated is continued over a period of months, indeed, many years. It is a stirring tussle, a resistance to the forces of nature with one end in view, namely, to wrest an upright posture from the compulsion of gravity's laws. This deep inborn aspiration is unerringly and irresistibly realized, to raise the head, one's own little earth sphere, towards the sun. The earthly substance in the sense organs, so conspicuously concentrated in the head, is thereby made so permeable and sensitive to light that it becomes a bridge

2.2 MAN AS MEDIATOR

for the soul that longs for the wide world. How much solid, dull, spherical material is rescued from the bonds of gravity and thus redeemed from the 'Fall' by generations of human beings all over the earth, raising themselves into an upright posture! No less wonderful is what is evoked by the other movement which confronts the first and penetrates it. And what takes place here is by no means 'more natural.'

Laboriously, man has also to learn the use of his limbs in early childhood. In the beginning they seem to be no more than appendages, somewhat like the feathers on an arrow that lie backward and follow the direction of flight. Here too, a reversal must take place. One can see it quite clearly in the small child that sleeps with his arms spread above his head. This position is a sign of health in the early period. Then, one day, he pulls his arms down and puts them under the blanket. With this action there grows a greater skill in terrestrial manipulation.

More and more, the limbs become tools; the legs enable one to move and reach a desired destination; the arms, hands and fingers mould the environment technically and artistically by all possible acquired abilities. Thus the momentum of the radiating cosmos-orientated limbs is directed towards the earth in which they form structures of light. How much humanly mediated wealth of light flows into the dark sod when, in a lifetime, a farmer cultivates a field! How much formative fullness of spirit is put into material when a craftsman skilfully and artistically works upon it! How much of the sun's metamorphic power streams incessantly into the life of our planet through the activity and transforming power of humanity!

Man's structural growth designates his mediating position in the world. The divided halves of the world are reunited in his being inasmuch as he raises the sphere of his weighted earthly forces towards the light and carries the winged sun, radiating from his hands, over the earth.

Does man perceive the mission marked out for him in the

world? Does he take hold of the ability to mediate that lies within him, or does he tear this increased double role apart? Where will he find a sustainer who will help him master the wide range of his being and use it as a controlled healing power to unite the world?

2.3 The sun-origin of love

It is useful, for man's self-knowledge, to be able to evaluate the spheres of influence of the opposing cosmic forces. When I know the influences I am subjected to, then perhaps I need not succumb to them. Many inclinations, compulsions and patterns of behaviour that otherwise would remain incomprehensible now acquire meaning. And what must appear to the uninitiated as sensual incitement or seduction can, when rightly understood, become an instrument for good under favourable circumstances. An obsession must not be immediately condemned as an 'aberration' or even as a 'sin'. It may be regarded as a loss of equilibrium that can be brought back into balance again by the right give-and-take in the art of living.

Let us look at human life from this standpoint. Look how often a small child knocks his head on corners and on the ground before he learns how to live with this hard 'object' and manipulate its size and weight? It shows how unaccustomed the soul is to the law of impenetrability appertaining to the world of physical matter. In the realm of the spirit from which it comes, such incompatibility does not exist. In the spirit, all beings interpenetrate one another, without colliding, and the first earthly experiences too reflect the characteristics of the spiritual life.

During the embryonic period, the child is *inside* the mother's body as a being that lives within another. From the point of view of life *on* earth, this is a remarkable oddity, almost miraculous.

But for a soul coming from the heights of the spirit it is the obvious continuation of the usual form of living together. Only with the expulsion from the womb is there an incisive change that demands a fundamental readjustment of the soul to totally different and contrasting circumstances. Suddenly, the law of 'indwelling' is no longer valid, instead, there is exclusiveness and the hard, unrelenting buffeting in three dimensions. A little newcomer from the land of the spirit must first laboriously learn this. And the skull plays its part in implementing this learning. From the beginning, it is the organ most equipped with earthly solidity. It can be used to gain practice in dealing with terrestrial gravity and material solidity.

The experience we derive through the limbs is quite different. In their structure, our hands and feet remind us all our lives that we are not exclusively citizens of the earth. In the midst of physical laws, they continually demonstrate the possibility of action contrary to the obstinate impenetrability of material substance. We must not then, on any account, withdraw ourselves into a purely inner realm when we desire to escape from earthly restrictions as Schiller seems to suggest in the lines:

> Thoughts live at ease together,
> But things in space hit each other hard.

Rather do we have in external life a sphere for exercising toleration and the ability to 'live at ease together' at any time.

'Things hit each other hard,' this applies inevitably to the head. The limbs, on the other hand, are used expressly to overcome repulsion and exclusiveness and to establish relationship. Caress, embrace, overcoming separating barriers, these are the words used to express behaviour of the limbs.

We greet each other from a distance by waving. When we come closer, each offers the other his hand. In shaking hands, both bodies are united for a moment in a harmonious whole. Encounter is expressed by gestures of inclusion. By touching one another,

2.3 THE SUN-ORIGIN OF LOVE

we attempt to establish friendly commitment and unity. It is a sign of particular intimacy when two people closely intertwine their limbs and remain wrapped together for a long time. Lovers walk arm in arm and with their arms round each other's waist. Holding hands is a particular sign of love. The extreme case is seen in the embrace which ultimately leads to sexual union. In this instance, one talks of 'cohabiting'. This is worthy of consideration. The archetype of 'cohabitation' presented by the great womb of the sun within which are the orbiting planets can never be perfectly attained on earth. However, there are different degrees of approximation in all the movements our limbs guide us to, from the tenderest caress to the most intimate embrace. As organs of the sun's rays they contain the riches of all the gifts of love that issue from the original source of loving embracement.

> Be embraced, you millions!
> I give this kiss to the whole world.
>   (Schiller, 'An die Freunde')

We must not forget the kiss! It is one of the most influential languages of love. Just because it works in such a mysterious way, it expresses the fire of the love relationship all the more poignantly. One could, indeed, object that the mouth belongs to the head. But are the lips not the 'limbs' of the head? The whole is again reflected in miniature: above, the actual skull-cap, encapsuled in bone; below, the mobile jaw that, with teeth, tongue and lips serves linguistic communication as well as the metabolic system. We kiss with the mouth and release thereby glowing impulses of love even from our earthy, solid pole.

Many old pictures testify to the affiliation of the limbs and metabolic life with the sun. The radiant power of the mother star shining over the world is shown with its midpoint and source coinciding with the lower pole of the human figure. There, where substances are annihilated and forces generated anew, where matrimony and birth, cohabitation and habitation have their

Figure 21. Il Sole, *the sun god, with his 'sun children' shown in their typical activities, fifteenth century.*

2.3 THE SUN-ORIGIN OF LOVE

cosmic centre, where there is no room for anything enduring, where everything merges into fiery fluctuation and creative cosmic renewal, there, is the *sun origin of love*.

(It is appropriate in such representations that various limb movements should be shown in the lower portion of the picture. This is a combination often made use of, sometimes supplemented by lovers and worshippers. Sport can be an enriching exercise in experiencing our limbs. However, there are limits that lie in the nature of the subject itself. How is it, for example, with boxing? Clenching the most delicate of our limbs and punching with the fists, is that not a perversion that converts the true nature of the body's limbs into its brutal opposite?)

2.4 More contrariety

Let us reconsider our cosmic opposing players. If, on this occasion, we direct our attention not to the structure but to the movement, then we notice a new aspect of their fundamental difference.

In rotating on its axis, the earth drags round with it a whirlpool of air and water. This, at any rate, is how we explain many of the ocean currents and air streams in the atmosphere. The lighter and more mobile layers lag behind the movement of the solid earth and there arise currents of air and water that run counter to the West-East rotation of the planet. This drag is greatest at the equator where the globe is thickest. There, we find the great equatorial ocean currents as well as the strong air streams known as the trade winds.

Anyone dressed in light flowing robes can demonstrate such a dragging current by spinning round. We could spin round on a revolving stool and notice how the light material twists as it lags behind the movement of the body. This natural inertia is often met with in everyday life, as when a rapid turn of the head makes our hair fly round.

Do we not already surmise that things are quite different with the sun? Indeed, there everything is again the opposite. The sun's orb also rotates about an axis, as can be inferred from the movement of the sun spots. But what is the nature of the movement in the outer layers? We observe the fastest rotation on the

2.4 MORE CONTRARIETY

equatorial belt, that is in the sun's 'tropical zones'. No hanging back is to be noted at the thickest part; on the contrary, the layers hurry on ahead.

Imagine you were spinning round on your revolving stool and your light garments, instead of flowing out behind you, were to fly ahead of you! Or how would you feel if your hair began to gyrate and you had to spin to keep up with it in a whirling dance? You would become dizzy; you would be afraid to 'lose your head'. It is not otherwise when we try to feel our way into the sun's rotation. What happens there makes us really dizzy. We do not perceive there any 'inertia', any opposition to rotation in the usual sense. On the contrary, if in the case of the earth the force of rotation appears strongest at the axis and falls off towards the outside, in the case of the sun, this is quite the opposite. The mobile force is most vehement outside on the tropical belt. The sun's axis, on the other hand, plays no active role; it does not motivate the whirlpool movement but participates in it on sufferance and reluctantly.

Is it surprising then, to find the moving force of the sun's rotation is derived from the surrounding space? For it has long since become clear to us that the essential sphere of the sun's life is in the extensive region of the planets. And in what genuine whirlpool would it be otherwise, be it the maelstrom in the sea, the tornado, the hurricane or the cyclone? The rotating movement always arises in the totality of the moving medium, and the eye in the middle, the hollow space, is ultimately a consequence of the rotation.

From a new approach, we arrive again at what we already knew. The sun's being is in its surroundings, all around. If we want to express our understanding of the sun aptly, we have the choice either of exchanging the terms 'inside' and 'outside' or of accepting the proposition that the sun is, properly speaking, 'outside itself'.

2.5 The being outside itself

Many people today seek to get outside themselves; seek ecstasy. The feeling is growing ever more general that in the present time we are all too hemmed in by our physical constitution. When we remain satisfied with our normal abilities we let the best that the world has to offer slip away. Our true nature withers in this spiritless, narrow civilization with its brutal deformities. We must get out, creep out of our skins, climb over barriers, widen our consciousness, attain to other soul conditions. We must get 'high' or 'be with it,' that is to say, we must be able to throw ourselves into the great wholeness and participate in the sum total of life.

> So that I know what holds the world
> Together in its most inward being,
> See all creative forces and seeds . . . (Goethe, *Faust I*)

The best people feel like this today, as they have done since the beginning of modern times.

To know means, at the same time, to go out of oneself. We shall not penetrate *into* the depths of the world unless we grow beyond ourselves. It is because we know this that we seek ecstasy. And we seek it even though we know the dangers of ecstacy. All sorts of means are being thoughtlessly used only to reach this goal; to break out of the 'damned, dull dead-end' and to fly up, out into the wide open spaces.

2.5 THE BEING OUTSIDE ITSELF

Everyone feels that this getting out of it and diving down is justified, but we all know what it means when we say of a person that, 'he is not himself' or 'he is beside himself.' It is known that a single dose of LSD may bring about a diseased disturbance of the mind that cannot be remedied. There may be more harmless poisons whose effect, at first, is not irrevocable. But who knows beforehand how susceptible he is? There may be robust constitutions that can stand a great deal, and they will not be satisfied with small doses at rare intervals. When they become addicted there is hardly ever a turning back and their personalities are deranged. Some may hope that in the future more efficient drugs will be discovered that will give access to the land of desires without evil after-effects, but it remains obvious that the taking of drugs is futile and leads to enslavement. For whoever makes his soul dependent on earthly substances will, in the end, attain the opposite of what he seeks. He will only exchange his boring dead-end for a magnificent, perhaps even enchanting, prison holding him fast in walls of material substance that are merely camouflaged. The addict, or he who can no longer live without medication, has deceived himself. He can no longer undo the spell of earthly substance; he can find no *freedom*. He gives himself up involuntarily to the thraldom he would escape from. it is a vicious circle. The path that promises to lead him out leads him ever deeper into the labyrinth of material imprisonment.

And the burning question is: is there a way out across the boundary that is not basically dangerous for human nature? Is it possible to augment the dimensions of the soul without instability or loss of freedom?

There is in the world *one* power, through which human beings can grow beyond themselves without losing themselves. This is *love*. Whoever loves, experiences a being-beyond-himself that is healthy. It is a process of fundamental transformation of self-love; self-renunciation and finding oneself at the same time. It is a change of consciousness that releases the grip of the narrow ego and yet does not estrange, but makes one at home with oneself and the world.

This open secret is known to all and yet we find it ever inexhaustibly new and desirable. The lover is more with the beloved than with himself and the feeling for the loved one is conferred on an ever wider circle. As Morgenstern expresses it:

> Maiden, will you be my world?
> World, will you be my spouse?

The 'I' finds its full reality only in this expanse of timeless harmony, in unity with the world.

> You are the eternal canopy
> You, my body, divinely given.

When the lover is 'out of himself' in the best sense, he feels his own being *heightened*. There is an extension of the soul's radius and an *inwardness turned towards the world* – this is love. Whoever has been in love knows how appropriate the words 'high' and 'in' are. Love is the path that leads further than any other: 'Love is patient and kind; love is not jealous or boastful; it is not arrogant . . . Love bears all things, believes all things, hopes all things, endures all things. Love never ends' (1 Cor. 13:4–8).

A true lover never loses self-control. On the contrary, it is through his love that he gains sovereign self-determination. Nor does he make himself dependent on external specifics from the dispensary of the earth's pharmacy. For the flame of love can only be kindled from within.

Therefore, love, and only love, can make man free.

Expansion of horizon, growth of soul and consciousness — unpredictable stages are conceivable on this path of love; stages on the way to a universal sun dimension that can survey a planetary system and uphold it. With this, we have arrived at the point which we had already reached in the first section, where we heard Christ speak in Jesus of Nazareth; 'I am the light of the world'.

2.5 THE BEING OUTSIDE ITSELF

The great sun-centred consciousness of an all-embracing cosmic leader rings in unison with this 'I'. We have arrived at the place where the reality of soul and sun coincide.

What the mother star of our cosmic system represents physically and what speaks from the loving, expansive heart of Jesus is an identical being. 'The Word became flesh and dwelt among us.' The sun Logos that let the earthly things become and grow as limbs of his being — 'all things were made through him' (John 1:14, 3) — has embodied himself in the form of man. And the ancient truth that the whole system of the heavenly spheres is his almighty divine sun-body, now rings transformed from human lips in Morgenstern's words:

> You are the eternal canopy
> You, my body, divinely given.

The flaming light of cosmic radiance, and blessed light of love, touches us in intimacy without blinding us. Through this power Christ, and only Christ, leads man in *freedom* along the path to the spirit.

> Not barr'd to man the world of spirit is;
> Thy sense is shut, thy heart is dead.
> Up, student, love — nor dread the bliss —
> Thy earthly breast in the morning-red!
>   (Goethe, *Faust I*)

2.6 Space and counter-space

We have inserted the chapter on conditions of the soul to make sure that our theme is anchored in matters that concern everyone directly. Our intention is that, when we now describe the fields of force between earth and sun in geometrical terms, we do not slip into academic abstraction but remain in touch with life. When, therefore, certain mathematical facts are put forward in what follows, the reader is requested to consider how these features, which he may find novel, relate to his own experience of *body* and *love* as the sphere of activity for earthly limitations and cosmic breadth.

The notion of two world conceptions is not foreign to mathematicians. They don't exactly relate their ideas to the sun and earth; indeed, when engaged in their higher mathematics they don't relate them to any external reality at all, for the world of pure thought offers them all the inner reality they need. However, they often find later, after having thought out their ideas long and deeply, that their conceptions do refer to the outer visible world. What was conceived in pure thought is seen to correspond to external reality. It is a reflection, a representation, of objectivity. Such a discovery gives rise to joy and surprise at the interrelating correspondence of what was intellectually formulated with the physical content of the world, as a die fits its imprint. We also find this occurring in the case of mathematical thinking respecting the truth of earth and sun.

2.6 SPACE AND COUNTER-SPACE

For the last two hundred years, mathematicians have been continually fascinated by contrary geometrical shapes and theorems that yet correspond to one another in pairs as image and counter-image. For example, it is a remarkable characteristic of points and lines that they can represent one another. Thinkers are continually returning in amazement to this relationship in their constructions, and to this day, so-called projective geometry remains a favourite study of mathematicians. But the question as to whether their discoveries in the realm of exact imagination has any relationship to objective reality is one which these meditating thinkers prefer not to ask. They do not doubt, however, that the Euclidean geometry most commonly taught in school adequately describes the shape and movement of objects found in our terrestrial world. As to whether the extended, doubly constructed projective geometry is likewise so related to reality, that its second half corresponds to an aspect of the world where it can be applied, this was of little concern to the discoverers. It was beautiful and more stimulating than any other branch of geometry and this was sufficient reason for their passionate devotion.

It was not until the beginning of this century that scientists took it upon themselves to consider seriously whether, corresponding to the concepts of theoretical geometry, there might not be an empirical reality which ought to be investigated. The Swiss mathematician Louis Locher-Ernst demonstrates in a new and particularly clear manner the dual nature of geometry so that one can see with one's own eyes to what extent our Euclidean view of the world represents only one half of reality. Locher-Ernst has succeeded in giving projective geometry a sound and unambiguous foundation, thereby making possible the practical application of these mathematical ideas. Instead of the word 'dual' Locher uses the term 'polar' to indicate that the duality is not arbitrary. The contrary pairs belong together and complement one another, thus expressing a genuine polarity. This Swiss mathematician is thus the founder of 'polar-Euclidean geometry.' This means that by using his techniques, we can bring complementary conceptions to the famous Euclid and everything arising through him during the past two thousand years. We can bring to every one of his theorems and configurations the polar counterpart. Locher-Ernst begun this work and in his book he used a style of presentation of striking clarity. Each page is divided down the centre into two columns in which the twin theorems and figures are placed side by side.*

Locher-Ernst admits that the content for the second column appears to be very unusual and strange at first. (For two thousand years we have been trained only in Euclidean ideas and it is only with effort that we can acquire contrary conceptions.) However, he challenges the scientists to familiarize themselves with the newly-discovered laws and to find out to which hitherto overlooked aspect of the world these ideas could be successfully applied.

In 1957 Locher-Ernst published a thoroughly comprehensive book which bears the character of a legacy because in 1962 the mathematician was fatally injured in the Swiss mountains. The title 'space and counter-space' gives the signal for research into a realm where spatial conceptions are contrary to those commonly accepted but are equally necessary for a grasp of the whole truth about the world.

The whole truth about the world can only be known when we learn to see the interaction of space and counterspace simultaneously. Therefore he quotes the famous French mathematician Michel Chasles who speaks of a universal dualism. Can one at all foresee where the consequences of such a principle of duality would end?. . . And can one say therefore that there is not, corresponding to the law of gravity, another law which plays the same role as Newton's and like it, would serve to explain celestial phenomena?

We should mention here that Locher-Ernst had a forerunner,

---

* This form of presentation was already used by T. Reye in 1866.

2.6 SPACE AND COUNTER-SPACE

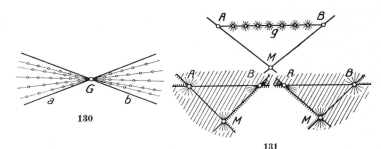

130 | 131

Winkel; deren Begrenzung wird gebildet von einem Segment der u. G. und zwei „*Halbstrahlen*". Ein Halbstrahl besteht aus den Punkten eines Segmentes. Wenn Verwechslungen naheliegen, sprechen wir ausführlicher vom *vollständigen Winkelfeld* und vom *gewöhnlichen Winkel.* Ein Winkelfeld besteht aus zwei Winkeln, solche Winkel heißen *Scheitelwinkel.*

Das Parallelogramm. Ein Vierseit, dessen Gegenseiten parallel sind, nennen wir Parallelogramm (Fig. 132).

Strecken; deren Begrenzung wird gebildet von einem Winkelfeld des a.M. und zwei „*Halbpunkten*". Ein Halbpunkt besteht aus den Geraden eines Winkelfeldes. Ein Segment besteht aus zwei Strecken, solche Strecken heißen *Scheitelstrecken.* Selbstverständlich verwenden wir das Wort Strecke auch im gewöhnlichen Sinne, wenn keine Verwechslungen möglich sind.

Das Zentrigramm. Ein Viereck, dessen Gegenecken zentriert sind, nennen wir Zentrigramm (Fig. 133).

Figure 22. A page from the book Projektive Geometrie *by Louis Locher-Ernst. Figures and text arranged in two columns.*

later a colleague, in George Adams only who already opens up a scientific view of the counter-space fields of force, through which we immediately see ourselves to be in unanimity with that which we have recognized as the sunlike activity in the cosmos. We shall return to George Adam's work in a later chapter. What does counter-space look like now? Let us look at the basic features.

2.7 An excursion into geometry

Here, we shall put things in such simple language that everyone can follow our thoughts. We cannot, of course, be expected to plumb all the depths of this beautiful science, geometry. However, we shall acquire some experience in a few subtleties that will begin to bring about radical changes in our feeling towards the world.

What is the simplest entity in geometry? Naturally, the point. This is what everyone will say, because he will think that everything can be constructed out of points, therefore they are the building stones, the elements. Is this really the case? Is this the sole truth?

Mathematicians know that this is not so. It is only *one* possible aspect of the world, constructions from points. What alternative can be imagined? Let us tell a little story. It may sound implausible to begin with, but it will help us to answer the question.

Two travellers meet on a train and enter into conversation. They soon discover their common interest in the art of structure and pass the time on their journey talking of geometry. It is immediately apparent that each approaches the subject from a very different angle. One of them is clearly quite a normal man of the world. He has obediently learnt his lessons at school and now tries to instruct the other. He, however, seems to come from an entirely different world from which he brings strange conceptions. Yet these are as consistent and convincing as those of the man who had a normal schooling. Thus through their contrary views there arises an astonishing give-and-take of knowledge and a broadening of outlook.

In order to clarify their words the two men confine themselves to plane figures which they sketch rapidly on small pieces of paper. The first man, the normal thinker (we shall call him N) maintains that the point is the building stone for all further construction and believes that he has thereby established a secure basis for all further discussion. But he quickly sees that he is mistaken when the other man (we shall call him O) replies: a point arises only where two lines cross. For me, straight lines are basic; a plane is filled with them as is the universe with the rays of the sun and the stars; and everywhere where lines intersect there arise 'points of intersection'.

Following this explanation, the conversation so develops that The partners can place their statements side by side:

N: In my view, the point is the basic element. If I take *two* points, then we can draw a line which *connects* them.

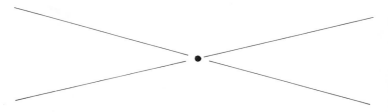

O: I perceive that what you are saying corresponds exactly to the opposite of what I have just said: If I take *two* straight lines, then a point is established where they *cut* one another. In my view, every point is a *point of intersection*, while for you, every straight line is a *connecting* line.

2.7 AN EXCURSION INTO GEOMETRY

N: Strictly speaking, the straight line consists only of points, like pearls strung out in a row, they form a *row of points*.

• •

O: I don't see only the two lines (or 'rays' as I like to call them) cutting each other in a point, but an abundance of them. Like a wheel with many spokes, the point presents itself as a *cluster of rays*.

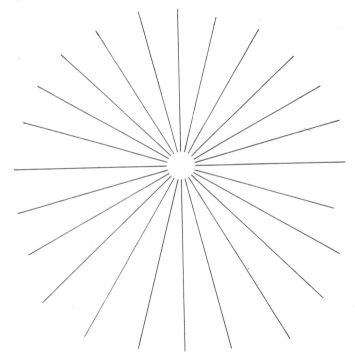

THE SUN

N: Let us now imagine our elements to be in motion. I think of the pearls, not as fixed on the string (as in a necklace) but as being *pushed along* . . .

O: . . . in a corresponding manner, my spoked wheel can *turn*!
N: My points *run* in a goose march, one behind the other.
O: And my rays *dance* like spokes round the hub.

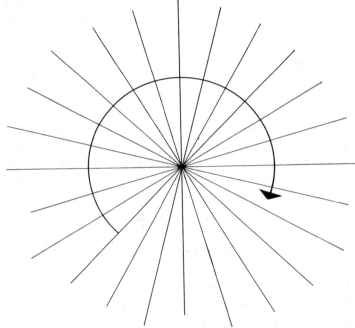

N: I perceive that you have arrived at the notion of *rotation*, as I have arrived at that of displacement or *translation*.

This confrontation is really interesting! And it is amazing how, approaching the matter from opposite directions, we yet

2.7 AN EXCURSION INTO GEOMETRY

possess an equivalent wealth of concepts so that whatever one of us says, the other can provide the corresponding concept in his own vocabulary.

O: It is indeed a peculiar kind of relationship between opposites. Let us try it a little further. I should like to see what happens when we limit the mobility of our elements. To begin with, my rays are not going to move unchecked about their axis, instead I am going to let one move round a short distance and then stop it. What do we see now?

N: A section of the whole cluster . . . because the wheel has revolved only a little way . . . now I see: the ray has moved through an *arc*. Now I shall put the same problem in my terms.

O: Let me try to think it out for myself. A point must not be allowed to move indefinitely in a straight line, but be stopped short. Between its start and its destination it will then have covered part of a straight line. I think you call that a *section*.

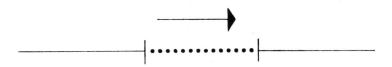

N: Exactly. You have learnt to transpose remarkably quickly! Now, everything has appeared so clear that I feel confident I could solve a similar problem. Please test me.

O: Then I shall set you a more difficult question. See now, in turning, my ray has traced an angle. But look at the rest of the cluster that it did not move through. What does that present us with?

THE SUN

N: Likewise, it is an angle! I think that is clear enough.

O: Quite right. But now conceive of the following: The ray could continue to turn and so move through the remaining area of the angle (2) until it arrived back at its starting position. (Strictly speaking, it would have to turn even further in order to reach the exact position it departed from. It would have to make another half circuit (of 180°) through (3) and (4), it would then have crossed both angle areas twice.) What would this movement correspond to in your scheme?

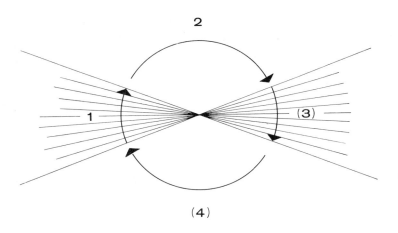

N: That is, indeed, a most peculiar affair. From your view of angle relationships it is quite clear. But how am I to express it in terms of my section? If a point had moved the whole length of the section (let us suppose from left to right) then if it continued its direction, it should eventually arrive at the place it started from. But this lies to the left! How, if it is travelling towards the right, can it ever arrive on the left side? (Furthermore, even then it should not really be at its starting

2.7 AN EXCURSION INTO GEOMETRY

position but should have had to travel along the line a second time.)

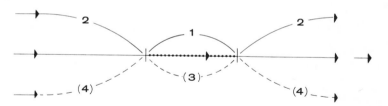

O: In fact, we have a riddle. You mustn't try to solve it just now, but ponder over it later at your leisure. At any rate, you are aware that your one-sided manner of observing figures has not only kept half of the world hidden from you, but even in that half you regard as your special domain, your knowledge remains defective as long as its contrary aspect is unclarified.

Well now, let us take a few steps further together. Perhaps you could make a note of the following questions for future consideration; If the whole cluster of rays through a point (the complete circle of spokes) is divided into two arcs by two rays* — could one not say that a straight line is divided into two sections by two points? One would never be able to see the whole of one of the sections because it runs towards the right and the left into infinity . . . But one could learn to see the straight line as enclosed within infinity! With regard to its passing through a second time, I should really like to leave that to your own investigation.

You see, the world is full of novelties and it is worth thinking everything through from the very beginning, even the most simple things. 'Unless you become like children . . .'

N: I know that all true science begins in unbiased wonder, and I have wondered much since coming into conversation with you.

---

* The angles, or arcs opposite each other are, of course, really two parts of *one* angle.

145

THE SUN

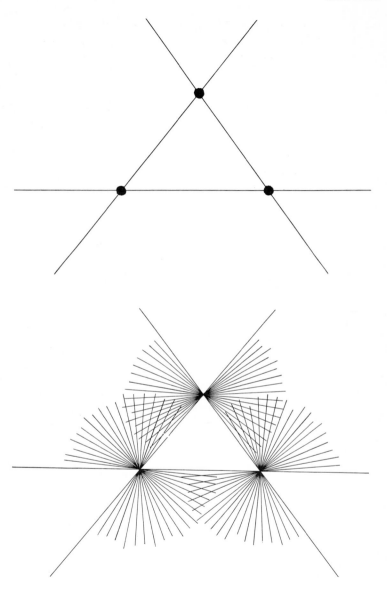

2.7 AN EXCURSION INTO GEOMETRY

Now I am curious to know whether you will throw any new light on the simplest figure in my school geometry book.

I take three points. Each pair can be connected by a line to form a triangular area. Strictly speaking, I must regard it as a *three-sided figure* . . .

O: As you already know, I would begin with three rays. Each pair cut in a point. The angles raying out from these points form a figure which I shall consequently call a *tri-angle*. But don't let me interrupt . . .

N: . . . With its characteristic lines, my three sided figure cuts out an area in the plane. This is the inside which is clearly defined over against the surrounding space outside.

O: Explain to me in more detail what you call the 'inside'!

N: In order to satisfy your preference for rays, I shall use as an aid a line (*t*): Imagine an arbitrary line through any point 'inside' the area, then one can say with certainty that it cuts *two* sides of the three-sided figure. (A little thought will show one that it is as impossible for it to cut one line only as to cut all three at once.) That must be clear!

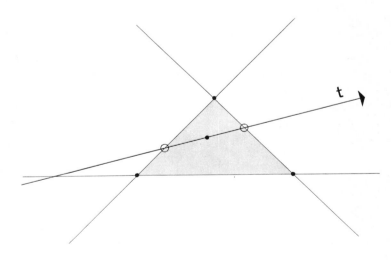

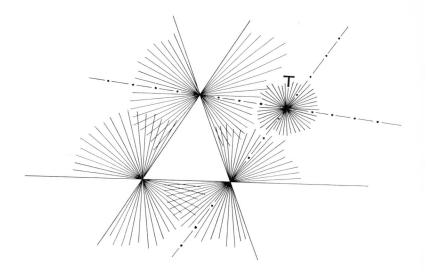

However, you asked so persistently about what I call the 'inside' that I suppose you must, again, have quite a different conception. Out with it, please.

O: Indeed, my approach produces exactly the opposite conception. My triangle also divides the plane into two clearly separate regions which one can call 'inside' and 'outside.' If now, I attempt to show, as you have done, which is the 'inside' then we have the following:

In deference to your liking for the point, I shall take as aid, a point (T). This point, considered as a cluster of rays, you remember, contains lines of the *inside* area only, and only then, when it comes to lie in exactly *two* arcs enclosing the three-sided figure.

N: I find that difficult.

O: Once more, then. The test line t could only pass through points (pearls) within the area if, on its way, it crossed *two* sides of the three-sided figure. So one can say with certainty of the

2.7 AN EXCURSION INTO GEOMETRY

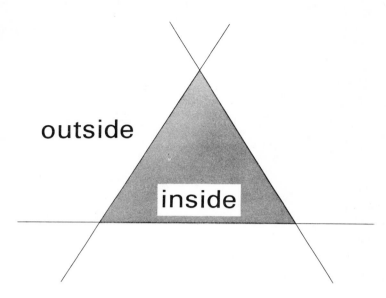

test point T that it lies in the arcs of two triangle points if two of its rays (spokes) also belong to the 'inside region' of the triangle. The corner points are like beacons that illuminate a specific arc. The test point is, therefore, illuminated by exactly two beacons (not only one, nor by all three). The little sketch will make that perfectly clear to you.

N: Yes, with some effort, one can think of it like that. But however I may regard the picture, I would have to change my present conceptions completely in order to arrive at an inner conviction. In fact, what you call 'inside' is for me the outside and you would regard my enclosed inner space as the 'outside' beyond it. The world is inside-out!

Here, we shall leave the travelling geometricians. One can imagine that, if they persist in the subject, they will bring to light some valuable and remarkable contrary relationships.

The unprejudiced observer who has followed the different arguments of both men will have learnt, however, that it is only in the

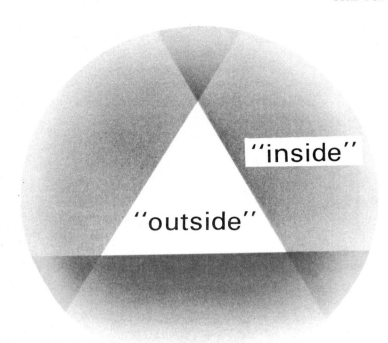

comprehension of both views that the whole of reality can be apprehended. And the conjecturing reader will have guessed that this duality in geometry is going to be a useful tool for exact mathematical treatment of the polarity 'earth-sun.'

We shall look at this a little more closely.

2.8 Kernel and husk

Those who have followed the conversation between N and O attentively will have noticed that both used the same logical form. They differed only in the use of words and concepts.

When N says:	then O uses the contrary expression:
point	ray (straight line)
two points connect	two rays cut
connecting line	point of intersection
row of points	cluster of rays
string of pearls	wheel of spokes
move along	rotate
run	dance
translation	rotation
section	arc
three-sided figure	tri-angle
inside	outside
outside	inside
region of points	realm of rays

The astute observer could say: If I learn the expressions which each side uses to meet the objection of the other, then, by using this conversion table, I can easily transpose the 'normal' geometry of N into the 'deviating' geometry of O. For example, instead of *point* one must always say *ray*; instead of *connect* one can insert *cut*, and so on.

We have taken quite a considerable number of word pairs out of the conversation. With the help of this table anyone could take a geometric proposition and, providing it did not contain a term absent from the table, could convert it into its 'opposite' without needing to understand anything in geometry. It is a simple linguistic rule.

Of course, the limits of this exercise are obvious. What happens in dealing with things that are not included in the list? Then, clearly, the conversion table is of no avail and new pairs of concepts must be thought out by mathematical imagination.

Let us use our initiative a little more! Having seen from the triangle that the spatial concepts of 'inside' and 'outside' can be understood in contrary ways, we shall have no difficulty in applying these concepts to other figures. Instead of a three-sided plane, let us take a circle! Without going through a chain of reasoning we can understand that, corresponding to the usual spatial inside of a circle there is a counter inner space which embraces what the first has excluded. And this can also be conceived of a sphere: a 'convex' (that is bulging) body of points can also be conceived of as a 'concave', hollowed out, space of rays. The latter figure surrounds a spherical space on all sides; the first fills it out.

Here, linguistic imagination can be brought to bear and, following the example of Locher-Ernst, the dual concepts 'kernel' and 'husk' introduced. Our table allows us to continue a little further:

circular region of points	circular realm of rays
spherical region of points	spherical realm of rays
convex (bulging)	concave (hollowed out)
filling out	surrounding
kernel	husk

2.8 KERNEL AND HUSK

Those last metaphorical words give us the key to what we are looking for: obviously, the polar form of kernel and husk reflects a relationship of the polar nature of earth and sun. We shall put it quite succinctly:

The *earth* is a *kernel* edifice. The compressed substance of its spherical interior may be regarded as an accumulation of material points, corresponding to the interior space of a convex region of points.

The *sun* is a *husk* edifice. The fullness of light and life spreading throughout the cosmic surroundings has its geometrical equivalent in the interwoven rays of a concave husk realm.

These two propositions present a programme of research for future astro- and geophysicists. Up till now, scientists have sought to understand the sun is restricted, one-sided, point-and-kernel thinking. They arrive therefore at cramped, alienating conceptions. This is understandable because their mode of investigation runs contrary to nature. Every subject of research demands its own appropriate method. As far as the sun is concerned this can only be found through an image counter to the conceptual devices suited to the earth.

We have seen in the last chapter how the new geometry provides us, in a wonderfully clear and unambiguous manner, with just such an instrument for an exact mathematical treatment of the sun's phenomena. Polar Euclidean counter-spatial conceptions offer astrophysicists a method which will introduce a new epoch to science. And one can expect the most informative and far-reaching results when this new direction in research is taken seriously.

The general nature of this book demands that we be satisfied with this hopeful outlook on future developments. Those who are interested in probing further into the more exact details of he mysteries of double space conceptions are recommended to read *Projective Geometry* by Olive Whicher, as well as those other books already mentioned.

2.9 The case of the pickled gherkin

Who could describe the astonishment of the explorer returning from his world travels to his homely bay, when suddenly he sees there the marvels of foreign places reflected on a small scale. Imagine the surprise of unexpectedly seeing a well-known face in some remote place.

It was at the beginning of autumn. The market stalls were laden with pickled gherkins. It is then that one thinks of the delicious piquant flavour of fine pickled gherkins, of old secret recipes and of the fat earthenware jars in which to store them. The thought of last year's delicacies makes the mouth water . . .

Now they are lying in spiced brine. In eight or ten days they should be ready, when their flesh will be soft and semi-transparent and they can be taken from the solution. Ten days seems a long time to wait! Could one not try savouring them earlier just for once? What difference will it make? Perhaps one of them is already tender and tasty?

The selected one is pilfered and laid on the table. The secret of its ripeness hides behind its slippery skin. A sharp knife will reveal it. Two clean sharp cuts and at each end a thin slice is taken. And what is revealed to the expectant gaze?

The salt solution had not yet fully penetrated the gherkin. On

Figures 23 and 24. Sections of pickled gherkin. Lighter parts show the different stages, yet unpenetrated, at stem and flower ends of the fruit.

2.9 THE CASE OF THE PICKLED GHERKIN

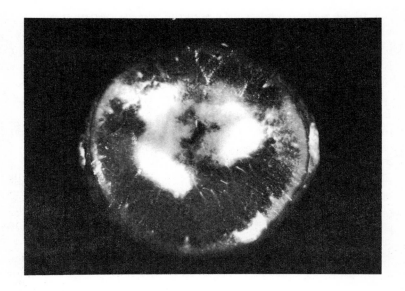

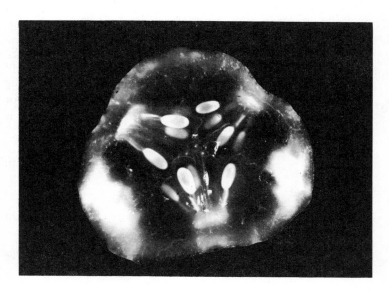

one of the sections the brine was in the process of penetrating the inside of the fruit. On the outside it was already semitransparent, but in the middle of the section there was an untouched circular region of greenish white flesh. This was to be expected.

But what met the astonished eye at the other end? Here, too, it could be clearly seen that the ripening process was not complete. But the green and white flesh was differently distributed. Indeed, the conditions were the opposite! It was the middle region that was fully penetrated, whilst in the outer circumference the hard unpenetrated places remained. Obviously, the process worked in the opposite direction. There, from outside inwards, here, from inside outwards. The result, in the first instance, was a hard centre in the middle of a soft semitransparent husk and in the second, a hard husk surrounding a semitransparent middle region.

The reader who has followed the sequence of thought in this book will immediately recognize in these polarities, the 'kernel' and 'husk' with reference to the fields of force of earth and sun. Thus the fruit mysteriously conceals a reflected image of the great cosmic relational tension in the penetration of the gherkin by the brine. At the stem end of the fruit there is an image of a firm earthlike centre; at the flower end the circumference appears as conditioned by the sun.

In this curious example, are we not touching the tip of a general law of the plant growing between sun and earth?

2.10 The plant between sun and earth

Sun – earth – plant. This is a trio that belongs so naturally and necessarily together. Its harmony surrounds us on all sides in endless repetition. Every time we experience it, we have a feeling of reverence towards all growing things. This is well illustrated in a simple little grace before meals by Christian Morgenstern.

> Earth who gave to us this food:
> Sun who made it ripe and good:
> Dear Sun
> Dear Earth
> By you we live.
> To you our loving thanks we give.

A third entity that connects the great cosmic polarity of these heavenly bodies is the 'plant being.' In its roots, fibres, woods and ability to transform chemical substance, the plant is fully a creation of the earth. At the same time, however, it is a creation of the sun; its green leaves express this as clearly as do the colours of its flowers and the sweetness of its fruit. Were there no sun, the earth would be bare of plants. Thus the plant, this wonderful living being, belongs simultaneously to the spheres of influence of both sun and earth. No barrier steps between. No darkening of the soul or estrangement, no constrictions or imprisonment in a self-centred inner life disturbs the clear interplay of forces between the two worlds. Where else is the interplay of sun and earth forces more clearly reflected than in plant growth? The human being whose mediant position in the cosmic polarity we have already considered, could learn much from the plants about his own behaviour. If he were able to acquire consciously, through his own effort, their innocent sincerity, he would have completed a truly great task in life.

> Do you seek the highest, the greatest?
> The plant can teach you.
> What it *is* involuntarily, you must *be* voluntarily . . .
> (Schiller 'Das Höchste')

In a comparison of man and plant we arrive at an understanding of a basic fact in the cosmos. Originally, man also stood in a mediating position, but to an even greater extent. However, he has lost his first translucent, receptive nature. Self-indulgence and wilfulness have obscured his perception. A separation has occurred; he has fallen asunder. This occurrence demonstrates what 'sin' is. We usually conceive of sin too moralistically and narrowly, using it to pass judgement on particular transgressions. If we understand 'sin' in its genuine meaning then we see it as a sundering (separation) arising from the emphasis on self. Briefly one could say: *Everything that differentiates man from the plants is sin.* In realizing this we are made aware both of the spiritual darkness in the soul and the promise and hope for the future given when freedom is recognized as mission. Both are in the meaning of sin. The possibility of error makes me sink into darkness but at the same time limits me towards freedom above the natural creation. I suffer from the sickness of sin and in overcoming it attain the highest form of salvation on earth. It is essential to understand that having left the innocent garden 'paradise' irrevocably behind me, there is no way 'back to Nature'. I must first 'make a journey round the world to see if there is not perhaps somewhere that is open again.' (Kleist, 'Marionetten Theater') What nature is involuntarily, you must be voluntarily.

2.10 THE PLANT BETWEEN SUN AND EARTH

The plant is the earth's *sense-organ* for the sun, or should we say the sun's sense-organ for the earth? An organ which perceives the duality of the cosmic forces and copies them in its own compliant nature. What for man is a goal to be consciously striven for, is visible in the forms of the plant kingdom in spontaneous perfection. It can be made the object of scientific research.

George Adams and Olive Whicher, mentioned earlier, wrote a book entitled *The Plant between Sun and Earth*. As the title indicates, the work deals thoroughly with the three aspects of our theme. The astonishing correspondence of plant form to the polarity of the sun-earth field of force is illustrated in many beautiful colour pictures.

If botany can be so augmented by a study of the sun, should not this new science, 'Plant between sun and earth', be a source of pioneering knowledge for solar research? Reading George Adams and Olive Whicher we gain this impression, and we can experience how isolated, specialist sciences begin to be united in a single science of man and the cosmos. Of their book, it is no exaggeration to say that the treatment of this triad leads the reader into a holy drama of initiation that brings heaven and earth into living dialogue.

2.11 Turn-about on the sun's rim

If plants can tell us so many interesting things about the encounter between the earth and sun field of force, we may wonder if there are not other places in space that manifest a living expression of this polarity. We might at first perhaps think of the opposite place, the rim of the sun's orb.

On the earth's skin, the plant kingdom, a collection of sensitive perceptors indicates what is happening. This zone is a frontier region. Here, the earth's interior meets sun-filled space. What is it like, then, on the 'opposite side', that is, where the 'sun's interior' of cosmic space meets the sun's skin? Is there anything there corresponding to the earthly plant kingdom? We leave this question unanswered here, as incitement to further research. It would, indeed, be rewarding to compare, in the light of the polarity of space, events in the different boundary regions of the sun with phenomena on the earth's surface; not only with plant growth but also, for example with the weather process, with butterfly and bird life.

From the distant region of the sun's rim, we obtain, in the first instance, experiences of *light*. These present us with a remarkable system which shows, as was to be expected, the contrariety of the spaces which contact each other. The astronomer, Thomas Schmidt (1968), describes what can be observed there, as follows:

> When, at the end of a total eclipse, the sun's disc appears from behind the moon, there emerges first, for a brief

2.11 TURN-ABOUT ON THE SUN'S RIM

moment, an intensely-coloured, thin sickle, the 'chromosphere,' which is the outermost 'skin' of the sun. Through the prism, it is seen to have a spectrum of pure lines of emission. Immediately thereafter, the full, natural white sunlight breaks out again; at the same time the previously shining lines of *emission* are turned into their opposite, the now dark lines of *absorption* . . .

What does this mean?

In order to probe deeper into the mystery of light sent us by the stars, astronomers have, for some time now, made use of a method called spectrum analysis. Here, a prism is fitted into the telescope and the light from a star is thereby transformed into a coloured band. No longer does each star appear as a point of light, but as a coloured streak, as though one had taken a tiny cross-section out of the rainbow. Each star is seen to have a different pattern in its 'spectrum', (the name given to the coloured band). In classifying the stars according to the nature of their spectral colours, scientists have made some revealing discoveries about the heavenly lights.

What is important for us is the fact that all the light phenomena of the night sky are governed by one universal law of polarity. It is just these remarkable circumstances that Thomas Schmidt deals with in his essay. He shows that what is observed on the sun's rim also fits in with the universal law. Two contrary types of spectrum images are observable. On the one hand, there are those in which single coloured lines shine brightly; in others, there are dark lines which in places annihilate the colours of the continuous rainbow band. (The first are called 'emission spectra' because, in the laboratory, they were observed to emit coloured light. The second were named 'absorption spectra' because the same kind of image is manifested when a light source is viewed through an incandescent gas emitting a line spectrum, whereby certain colours are toned down or swallowed up.)

In the paper quoted, the astronomer considers the two

THE SUN

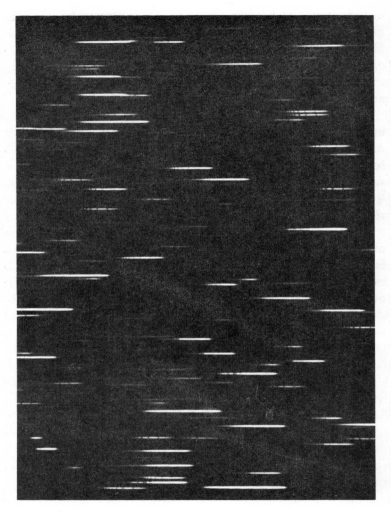

Figure 25. Star spectrum. The different patterns running through the horizontal spectral lines can be clearly recognized in the narrow bands of light.

2.11 TURN-ABOUT ON THE SUN'S RIM

occurrences of emission and absorption and the turning of one into the other. It is clear that we are again on the trail of a polarity. Without involving ourselves further in the physical aspect of the matter, we can ask: How does observation characterize this turn-about on the sun's rim, on the boundary between sun interior and sun exterior?

We shall follow the scientist in his sober observations leading up to his statement regarding the sun: 'Confining ourselves to a description of the phenomenon, we must affirm then: the *positive* substantial light effects which appear in the circumferance of the sun in the chromosphere, turn back into the sunlight itself (that is into the light issuing from the disc) in a *negative* substantial reaction'.

Does this turning of light into shadow not remind us of the exchange that takes place in the gherkin between outside and inside? In both events, is it not the same 'crossing over' of fields of force that occurs on the one hand in the realm of light, on the other in the substance of the plant's form?

2.12 Eighty-eight

When the external husk of the cosmic circumference and the earth's compressed sphere encounter one another as rays of force then a figure is formed in the middle of which is the cross. The two curves from above and below are united where they cross in a lineal configuration: the figure of eight or lemniscate.

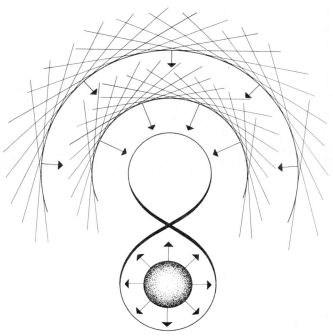

2.12 EIGHTY-EIGHT

Figure 26. Dandelion. Figure-of-eight crossing over between root and foliage.

The image of the 'crossing over' directs our attention once again to George Adams and his wonderful plant geometry in which the crossing of lines or planes is of particular significance. It happens just at the point where opposites come together. It is characteristic of plant growth that the cross figure is clearly delineated at all those places where sun and earth forces meet and penetrate one another.

Lemniscate figures permeate the whole of the plant form. One can say that the space round the earth where plants grow is full of living 'lemniscate spatial patterns.' Here, the medial position of plant nature is graphically expressed.

The crosslike loop of the growing points is most obvious where root and stem meet. This point on the stem below the first leaves (called hypocotyl in botany) lies indeed on the boundary surface of the earth. Thus there must be an encounter there between earth and sun influences. In fact, botanists have discovered that there is a complete crossing over of all the fibres at this locality. What was inside the root is on the outside of the stem (we know this well enough from the yellow core of the carrot) and what forms the outer layers of the root becomes the middle part of the stem. We have here an exchange between outside and inside round the axis of the stem! One cannot imagine a more perfect lemniscate crossing. Here, concretely, space and counter-space are united in the amalgam of pliable plant substance.

This first 'knot' which most emphatically demonstrates the interpenetration of the counter forces, is continued up the plant stem. Though it is not so obvious at first sight, nevertheless, further investigation shows us that all these nodal points are, in fact, lemniscate transversals. We find the same figure transformed in different ways all the way up the stem. Finally, at the transition to the flower the spatial interweaving is again clearly visible. But even in the leaves and the different parts of the flower we see subtle indications of the lemniscates expressing the earth-sun harmony that vibrates in all living growth. This is all graphically presented in *The Plant between Sun and Earth*, illustrated with drawings and excellent colour plates of the many ways in which the encounter with primordial space can be experienced. The forces moulding the lemniscate curves can be followed right into the formation of the fruit. To this classical work, we might add

2.12 EIGHTY-EIGHT

Figure 27. Potentized figure-of-eight formations along the plant stem.

our little chapter on 'the case of the pickled gherkin,' as an appendix. It would contribute another little stone to the mosaic of the overall picture of the plant between sun and earth and, perhaps, encourage others to make their own journeys of exploration into a realm only recently opened up.

THE SUN

Figure 28. The 'Eighty-Eight' (species callicore*). Butterfly from South America.*

Is this great insight not an incentive to further research into all the manifold aspects of the subject? Botany appears as the first ladder whose rungs we can climb with certainty and influence into the many-sided mysteries of the sun-earth continuum. On the other side, the reversal on the sun's rim comes to us as a counter movement. Does this not challenge us to seek other modifications of the lemniscate phenomenon in the rest of space, particularly in the intervening realm, along the whole line from the earth's kernel to the sun's centre? What a stimulus this could be for meteorology and research into the stratosphere and beyond, not to mention zoology, anthropology and, of course, astronomy and solar physics.

2.12 EIGHTY-EIGHT

As a testimony to all this, there appears in the midest of sun-filled terrestrial space, a delicate living creature that carries a prophetic message in its delicately swinging movements. This is a butterfly which bears the symbol of the lemniscate on its wings. It is written twice on each wing as though to confirm beyond doubt that our whole world is permeated by this wonderful law and it could almost be given the name of 'Eighty-eight'.

2.13 The sacred loop

We have considered a most mysterious figure. It was doubly emphasized on the butterfly's wings; hidden, yet perceptible in the lines of the plant's tissue and formative forces; translated into the language of light through the spectrum showing the conversion of light into darkness on the sun's rim. In the picture of the crossed, interwoven lines we are shown ever again the touching, uniting, penetration of opposite fields of force. Indeed, it appears to be *the* classical sign of such encounters of polar activity.

The peculiarity of the lemniscate is not unknown to geometry. If one follows the sequences through which the figure of eight loop develops, one observes an occurrence whose close relationship to the dual sun-earth world is immediately seen. In order to make the geometrical description easier to understand let us imagine it as the movement of currents in water.

Imagine a spring in the middle of a lake gushing up on the bottom. Its water will be distributed equally to all sides. We see the currents caused by a spreading out from the source. (Sometimes one can see the same thing happening in a cauldron during the making of jam when the liquid over the flame in the centre wells up in a 'heap' and flows out towards the sides.) To discover how far the currents of water travelled in a given time, a series of photographs could show a number of rings round the same centre (concentric circles).

There is also the contrary phenomenon. In the middle of a lake

2.13 THE SACRED LOOP

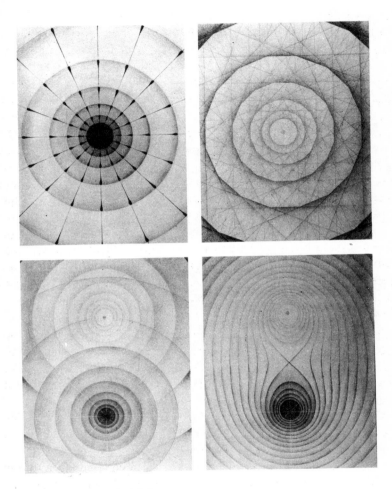

Figure 29. Spring and sink interpenetrate.

THE SUN

there could also be a place where the water disappears. When the water leaves the lake's basin through a drain, there will be a movement of water towards the sucking centre. Here, as opposed to the 'heap' noticed earlier, there will be a trough, and this 'counter-spring' is called, appropriately, a 'sink'. Again, a series of still photos of the currents in the water would show a number of concentric circles. The picture is the same in both cases. Only those who knew what caused them would know that two opposing forces are at work, that inside and outside have been changed about.

Bearing these opposing fields of current in mind, we can now ask: What happens when two such polar events penetrate one another; that is when in the same lake the spring and the sink function in each other's neighbourhood? What sort of current lines arise when the two fields of force meet in space? (Since we are here only concerned with the measurement of geometric growth, may we mention for the sake of the mathematicians among our readers that we are assuming the forces exerted in both centres are equal.)

Depending on whether the springs or the sink first began to function, difference curves will arise showing all the characteristics of the process of polar penetration. What do they look like?

Among the recently so popular sun-druses found in Brazil, there are some unique specimens that have retained the shape of such currents fossilized in mineral form.

The outer lines are oval, ellipse shaped. In the middle of the ellipse, between spring and sink, the inner lines bend more and more inwards towards each other until eventually they cross over and become two circles; the pattern is cut in halves. Now there are two enclosed islands which belong symmetrically together as a pair. Between both types of curvature, the uniform and the dually divided, a whole series of transformations is represented.

During the moment of its crossing over, while the islands are still laced together and before it becomes two isolated circles, the

2.13 THE SACRED LOOP

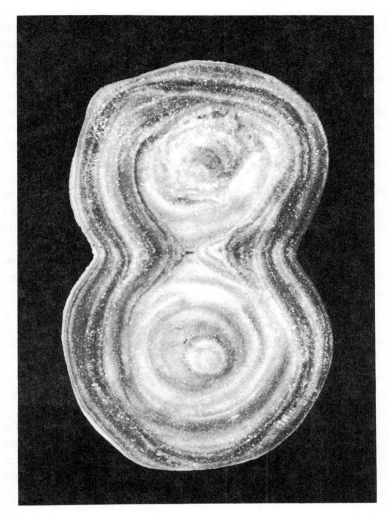

Figure 30. Sun druse from Brazil.

line reaches a transitional form. It shines forth in its passing. It happens only once, it is unique in the family of curves. In its form it also takes up a medial position. Between the oval that encloses both poles and the separated egg-shaped lines, stands the *lemniscate* as a clear two-limbed figure which however, with one comforting gesture swings back into itself, as much as to say that the bridging of opposites and finding their common meeting place is by no means unrealizable. (In the illustration of the crystal, the transitional form could not be shown. Here, where the lines of the figure eight cross over, there is a hole.)

The mathematical regularity of this family of curves was discovered by the brilliant astronomer Domenico Cassini (1625–1712). They are also called, therefore, 'Cassinian curves.' It did not occur to Cassini to investigate the profounder aspects of these figures as we, on the basis of modern geometrical insight, are able to do (see, for instance, Locher-Ernst 1938). Nevertheless, he toyed with an idea whose content lay beyond the scientific knowledge of Copernicus. Namely, that his curves, particularly that of the lemniscate, played an important role in the movement of the heavenly bodies. Cassini's surmise is not at all strange to us on account of our sun-earth observations. One could well imagine that the movement of the planets within the field of double tension (between the cosmic spring and sink, as it were) include further stages of intermedial lemniscates that are reflected in plants, butterflies and configurations on the sun's rim.

There are two ways in which conventional astronomy has imposed upon us an unsatisfactory world picture. On the one hand it presents us with a view of the sun as a fiery, terrifying ball, distant from the earth, and thereby contributes to the notion of the separateness of the planets. It sees the duality after the manner of Cassini, as egg-shaped islands, unconnected. Yet it obscures the otherness by measuring solar occurrences purely by terrestrial standards. Thus it builds up a materialistic cosmic cauldron that corresponds to the Cassini oval embracing the poles.

2.13 THE SACRED LOOP

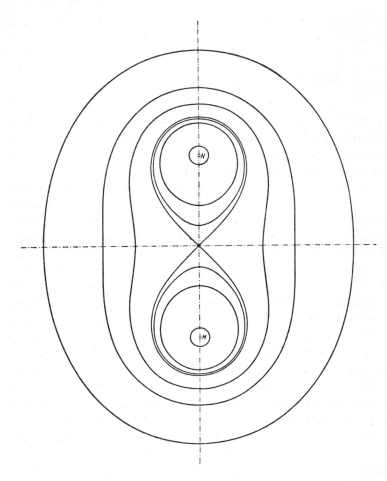

That there is a middle road that unites without combining, and separates without tearing apart is barely thought of in scientific circles today.

More profound investigators, however, take their stand in a realm where the mystery of the lemniscate impinges upon them and, like wise men of earlier times, they begin to apply it as a guide-line to the life of man and the world. As the poetess Nelly Sachs writes:

> I am in a strange country
> that is guarded by the 8
> the holy angel of the loop . . .

To be estranged is, indeed, to be without people. From the dreadful inner godlessness of existentialism to the arid sun darkening of solar physics, nothing is spared us. Nevertheless, this division in the world is a good sign. Properly understood, it can be a blessing — indeed, a protection. One can sense the force as a being that lets the tension between opposites merge in progressive motion — an angelic being, holy.

Can we turn to this being from within the lemniscate's sweeping curve? Can we turn to this angel to find the power to cross over, the power that builds bridges, that is itself the road between here and yonder, that is contingent upon itself and therefore can never lose itself? The power that takes up the cross of earthly presence and rounds it off in the encounter with cosmic otherness. Whom do we meet in the figure of the 'holy angel of the loop' that is ever on the way?

2.14 Christ in you

Christ is represented with unmistakable clarity in lemniscate form by the unknown painter of the frescoes in the chapel of Chillon Castle on Lake Geneva. In the language of the painter, could one desire a clearer presentation of the moving cosmic truth about the earth and the sun, their separation and their reunion? This picture contains everything that could be said in this book.

Is there an indication of more than the sacred lineal curves to correspond to the being of Christ as mediator between heaven and earth? The polar being of this dual figure, sun-god and brother of mankind, is woven into the figure of eight as the unifying seal of the human god. It appears as a spiritual necessity that the lemniscate is portrayed at the altar, where the meeting of the heavenly and earthly is celebrated.

Wells Cathedral answers in the language of a monumental architecture whose success was perhaps arrived at unconsciously, though it represents a spontaneous cultic impetus never again attained. Where the nave and the choir meet there towers in arched stone a mighty figure of eight. This is the point in the cathedral where 'counter worlds' touch one another, where preaching and the good will of a receptive congregation meet in unity and where earlier times placed a separating rood-screen. Here, the limbs of the Gothic arch of the roof come down to rest on a reflection of itself, an upturned arch, whose limbs cross at the apex and sweep down to the floor at either side of the nave, forming a further great

Figure 31. The fresco of 'Master Jacob' from the chapel consecrated to St George in the castle of the Counts of Savoy in Chillon, Switzerland, thirteenth century.

arch. This is the lemniscate of Wells. The topmost loop consists of two pointed arches, one coming down from the roof, standing on its inverted reflection. The lower loop is open, or rather, through necessity, closed by the horizontal floor.

The features of this great architectural structure affect the observer with striking conviction. One can get this impression even from the illustration. The focal point in God's house is delineated with strict accuracy of purpose, in such a manner,

2.14 CHRIST IN YOU

Figure 32. The lemniscate arches in Wells Cathedral, Somerset, thirteenth century.

however, that the door is wide open to the greatest penetration of the current that streams back and forth, as barrier and bridge in the unity of being.

We shall pass over the baroque builders (in particular Balthasar Neumann) who attempted to abolish the mystery of the church's spatial dimensions so that they were no longer outwardly architecturally symbolic but were left for inner surmise. Instead we shall direct our attention to the exponent of the cult, the priest. His sphere of activity lies just in that place where cosmic worlds encounter one another. Since ancient times he has been called *pontifex* (bridge builder), and his true priestly symbol, the crossed stole, is a lemniscate. It is open below — like the spatial configuration in Wells. Thus the continually enduring activity of the priest is perceptible. The open-ended lower half of the loop confers upon the symbol an inward direction, an exertion from above towards those yet unfulfilled below. The power of the mediator, here called upon to act, is directed towards the earth. Christianity is the 'faith that loves the earth.'

In the contemplation of cultic emblems, we are initiated into the conscious operation of the symbol. The lemniscate becomes literally a key that opens up adjacent realms and reconciles them.

We meet with an intensified form in the modern structure of the cult as practised today in The Christian Community. When the priest goes up to the altar before the gathered congregation he wears the ceremonial garment, the chasuble, on the back of which is an open U-figure, visible to the congregation during the greater part of the ceremony. The community of human beings comes together opening themselves to the higher world. During the procedure they experience the presence of the Other. Every time the priest turns round to bless the congregation by saying that Christ is in each soul, the sacred figure of eight appears on the front of the garment. It appears only for a brief moment. But as the service develops, it is seen again and again and the blessing is imprinted in the soul. Finally, at the climax, when the

2.14 CHRIST IN YOU

Communion is administered, the brief moment becomes an enduring period of time during which each communicant stands close to the figure of eight as bread and wine and the peace greeting are received.

What does the transformation of the U-figure into the figure of eight signify? Where in the former case the line was open, in the latter it is closed. The congregation, gathered together for willing sacrifice, receive an answer to their searching question. But the line is immediately intertwined; a cross appears in the middle, as though it wanted to feel itself, to become self-conscious.

If it were only a question of attaining to new *knowledge*, then it would suffice for the broken, open-ended figure, to close. That would be a good picture for the perfecting of knowledge, to turn the incomplete curve into a full circle.

In Christ's penetration, however, a *transformation* takes place. The answer penetrates the being of the questioner and makes him into another. 'Not I' says the U-figure in free, devoted sincerity. 'But Christ,' says the completion of the circle. 'But not Christ as a divine power, external to humanity, but "Christ in me",' thus speaks the cross as it becomes self-conscious.

Thus the bridge is built anew. Sun and earth, the heavenly bodies that had fallen apart, are brought together again in the symbol of the holy looped line. In Christ our planets are united.

>   . . . sun's weaving
>   love beams of a world
>   of creative beings
>
>   that through aeons of time
>   bind us to their hearts
>   and give to us at last
>
>   their greatest spirit
>   in human form during three
>   years: when he came into

> his father's heritance — now the earth's
> innermost heavenly fire:
> that it too shall once become sun.
>   (Morgenstern, 'Licht ist Liebe')

Practice in becoming conscious of this cosmic fact and entry into the healing double field of force in the act at the altar is, for human beings, synonomous with coming to oneself. In the 'consecration' (that is, perfection) of the genuine human being, the son of earth fulfills his disposition. He is destined as mediator between heaven's being and the being of the earth. For this reason, the cult in the divine worship of The Christian Community is truly and appropriately called the 'Act of Consecration of Man'. Could we desire anything more beautiful or greater than to participate, through conscious activity, in bringing about the perfection of the cosmos and transformation of souls out of the power of the 'Christ in Us'?

3

The Kingdom of Light

This sun mystery was experienced as humanity's greatest jewel. Under the Emperor Constantine, Christianity adopted an attitude which denied the sun.

Through a modern knowledge of nature, the sun jewel must again be made to shine.

Rudolf Steiner

3.1 Sunspots

At the beginning of January 1971, a fairly large sunspot was clearly visible to the naked eye. On wintry mornings, the disc glows pale and reddish through the misty air and can be observed during a walk over the snowy fields. Suddenly, one stops in fear: Was that an impurity, a black spot on the bright round disc? Or, more probably, an illusion? One moves backwards, forwards, sideways, looks away and tests it again. If the vision remains, then the observer is compelled to acknowledge that, indeed, a dark disturbance has taken place on the clear disc.

A sunspot! Imagine the feeling of the first discoverer who saw with dismay this blemish on the pure background and who admitted, after a long inner struggle, that indeed, 'the eye of the world can become sick,' that in the centre of the light, traces of darkening forces must be recognized.

Mankind in general became aware of sunspots only relatively late. Even the Chinese, to whom we owe the first reports and whose ancient culture made many discoveries surprisingly early, only began to write reports in AD 301. These were kept regularly until 1205. Did such sunspots occur in earlier millenia as in more recent times? What has been reported of the Inca, Huyana Capac, who lived around 1500 in Peru is remarkable. The discovery of sunspots caused such disturbance in his mind that he began to doubt whether the sun was in truth the Godhead. This is a shattering blow to the most fundamental belief of an Inca, who

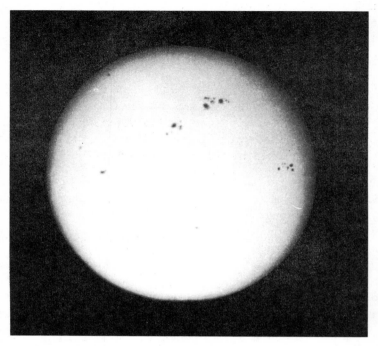

Figure 33. Sun spots photographed from the solar observatory, Wendelstein, on November 9, 1956.

embodies in himself the son of the sun, representative of the radiant God on earth.

The sober scholars of Central Europe, trained in the sciences, also fought against this observation, forcing itself upon them. Kepler interpreted what he saw as clouds passing in front of the disc, or perhaps other planets, Mercury or Venus. On no account could these dark elements belong to the sun-being itself. Johannes Fabricius, who used a telescope for his research on the sun, gives an entertaining report on his observations and his original hesi-

3.1 SUNSPOTS

tation to admit as true what he surmised. He describes what happened on a December morning 1610 (Wolf [n.d.]):

> I trained my telescope on the sun . . . as I observed it attentively, a blackish spot of no mean size in relation to the body of the sun, presented itself to me . . . I thought that passing clouds constituted this blemish. I repeated the observation perhaps ten times through Batavian [Dutch] telescopes of different sizes, and made sure that these spots were not caused by clouds. Yet I did not want to trust myself alone and called my father . . . We both began to catch the rays of the sun in the telescope; first at the periphery, then we went towards the centre, until the eye was accustomed to the rays and we could see the whole disc. Then we saw the object clearer and more accurately . . . So the first day passed and the night was a hindrance to our curiosity, which gave way to doubts . . . as to whether the spots were inside or outside the sun . . . The following morning, the spots appeared again, to my satisfaction, because I held the first opinion to be the correct one . . . but the spots seemed to have changed their places a little, which worried us . . . Three dull days followed. When the sky was clear again, the spots had moved diagonally from east to west. We noticed another, smaller spot on the edge which followed the larger ones, and moved within a few days into the centre of the disc. Another appeared and we saw all three. Gradually, the larger one moved beyond our vision and we could see that the others had the same intention. A kind of hope made me expect their return. After ten days the largest appeared on the eastern rim and when it had moved further into the sun, the others followed. They always appeared unclear near the rim. This led me to the idea of the spots as rotating.

Galileo observed sunspots at the same time, but had considerable difficulty in bringing his discoveries to the attention of others.

The Jesuit, Christoph Scheiner of Ingolstadt, examined and described the spots most thoroughly and was therefore rightly or wrongly celebrated as the discoverer. He was scolded by his superior for having seen something not contained in Aristotle. But Scheiner was not discouraged and continued his research. Soon, he calculated the speed of the sun's rotation about its axis by the movement of the spots, and observed finer details that found an accurate and comprehensive description very much later in the so-called butterfly diagram.

An interest in the observation of the sun and its sunspots grew up at the beginning of modern times, but it was only in the middle of the nineteenth century that a scientist collected all the diverse results spread over the whole world and arrived at a complete picture. The Swiss astronomer, Rudolf Wolf, gained distinction by co-ordinating the data of all his predecessors and adding his own research. He succeeded in formulating well-documented explanations of the origin and development of sunspots, and he is regarded as the pioneer of modern sunspot research.

It was his material that provided evidence for a theory others had surmised before him: the activity of sunspots underwent a regular fluctuation, so that periods with few spots (minima) alternated with those of many (maxima). He discovered a period of eleven years.

It was remarkable that around the time (1852) when he published his findings, other scientists (Gauss, Ørsted and others) discovered a corresponding fluctuation in the magnetic field of the earth. Wolf received the news with great excitement and, comparing the periodicity, found an exact coincidence. The frequency of the northern lights fitted into the same time pattern, as well as the occurrence of earthquakes, and changes in the weather. One can understand the elation that gripped scholars who had observed such fundamental regularities. Close co-operation was needed in many different fields of study.

Is it obvious that the origin of sunspots is not an isolated occur-

3.1 SUNSPOTS

rence, confined to the sun, any more than the change in the weather, atmospheric magnetism, the northern lights or the movement of the earth's crust are isolated independent phenomena; but that the fluctuations in all these different realms can be referred to a common cause, that in all these members of the universe, however far removed from each other they may seem, a common pulse is beating, the rhythm of a holistic planetary organism.

3.2 Symptoms on the skin

It is strange to see, unprepared, the dark spots on the radiant sun disc. Anyone who has studied enlarged photographs of sunspots or drawings by astronomers will admit that he is deeply disturbed by the sight. What we perceive is uncanny. The bright surface is damaged, brutally ripped apart, partially perforated, as if it bore festering wounds. The rims are frayed like suppurative, infested tissue (doctors call it 'necrotic'); it decomposes and looks blackish-brown, as if burnt. One is reminded of boils in rotting tissue.

The sight of wounds in a human body can call forth either disgust or sympathy. Likewise, the sunspots can frighten and repel or, through profound shock, lead to superhuman compassion for the cosmic suffering that appears before our eyes. We should not dispel this profound impression with hasty explanations. The way we have gradually learnt to regard the sun in this book leads us towards an understanding of this mysterious phenomenon of sunspots.

We are inside the radiant body of the sun, containing the whole cosmos and its planets, and we look on to the circular disc, on to the 'skin' that divides this inner sun from the outer space beyond. The direction of our gaze is, therefore, the opposite of one observing the human body. In the former case we look from the inside outwards, in the latter from the outside in. In each case, our gaze is interrupted by the dividing skin which lies between inside and outside.

3.2 SYMPTOMS ON THE SKIN

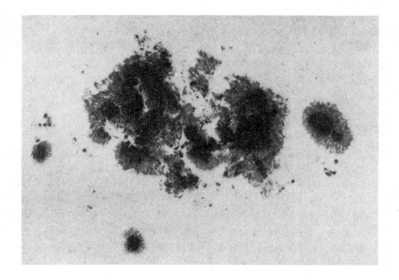

Figure 34. A group of sun spots, greatly enlarged. Photograph of the Mount Wilson and Palomar Observatory.

The skin forms a barrier, and yet at the same time it is a mediator, an organ of contact between different realms. It hides and reveals. An expert can, by its condition, decipher many things not visible to the eye. Indeed, the surface is a faithful mirror of the conditions within the body.

A doctor always turns his attention to the condition of the skin. He tells the patient with a yellow, flabby skin that he suffers from a disturbance of the liver and he administers immediate aid to the heart and circulation when a man's face turns blue. Different skin diseases or ailing hair and nails indicate an imbalance inside the body. Even the joys and sorrows of the soul betray themselves by blushing or growing pale, sometimes to our consternation.

Ought we not develop a sensibility as wide as the whole cosmos, a kind of medical observation, trained on the universe, in order to decipher the sign language of sunspots?

If we train ourselves in attentiveness, just as a doctor does who enters with sympathetic understanding into the unique individuality of his patients, then those dark cyphers may become the key to the life processes in the body of the mother sun. Research into sunspots must grow into a wider sicence of skin symptoms, obtaining results from descriptions of every change in the sun's surface to diagnose health and sickness in the solar system.

3.3 *Qui tollit peccatum mundi*

What is left to us except emotion when faced with a process that includes our own life as citizens of the earth and of this universe? Do we now have a better understanding of why, somewhere in our soul, we are deeply moved when we see sunspots? Indeed, this spontaneous reaction is based on the whole configuration of the world. The darkening of the source of light is not just a matter concerning a fiery ball millions of miles away, something of fascination only for experts, but the traces of an event influenced by our own behaviour — we, as planet beings, co-determining it. One has become accustomed to talking of sunspots as gigantic 'magnetic freezing machines' and, though it is an ugly and brutal expression, it can bring the truth straight to our hearts. Powerful processes of cooling and darkening, icy cramps, are occurring in the pure organ of the centre, documenting what happens round about in the community of the planets. The sun holds a mirror up to its children, its partners.

Observers have been occupied with the phenomenon of fluctuations in the frequency of sunspots. Rudolf Wolf, as already mentioned, using his comprehensive knowledge, estimated the regularlity of the fluctuations as a pattern of repetition within eleven years. His supposition that these 'periods' could be explained by other regular oscillations in the life of the planets has been neglected by later scientists, because it remained unproven. And so the eleven-year period stood for long as a

THE SUN

Figure 35. Group of dual polar sun spots. The whirlpool structure round both centres can be clearly seen.

concept in research into sunspots without any deeper insight into its origin. The phenomenon which drew everybody into its spell remained a riddle.

This scarcely changed even when a new characteristic of the sunspots was discovered at the beginning of our century: that their occurrence is ruled by a kind of double effect. Around the centres of the spots are radiations like whirlpools, rotating to right or left. Frequently the spots occur in pairs, in such a way that the whorls show opposite movements; for example, the first spot (in its movement across the disc of the sun) shows whorls moving clockwise and the second shows anticlockwise motion. Eleven years later, spots are found with the opposite pattern of movement and only after another eleven years does the original relationship return. So we conclude, according to these observations, that the sunspot

3.3 QUI TOLLIT PECCATUM MUNDI

period is, in fact, twenty-two years. Yet, doubling the number of years does not solve the riddle of this phenomenon. What is it that makes eleven and twenty-two key figures in the life of the sun? If we look at the sun in the way customary today, the question of a causal connection with it remains unanswered.

Dissatisfied with this vagueness, modern scientists move away from the hypothesis of a periodic movement. They look at each increase or decrease in the activity of sunspots as an independent phenomenon to which they apportion a period of eleven years. This only represents another extreme.

Neither the thesis of a rigid repetition belonging to the sun alone, nor that of one phenomenon following the next without connection, can, in the end reveal anything. The heart of the problem lies on our conception of the sun. From the point of view taken here, it is self-evident that the origin of sunspots is linked with a process of life and suffering within the whole system, and that therefore we have to consider the rhythm of numbers in the correlation of the planets themselves.

It is encouraging to see an attempt of this kind of understanding in the work of some scientists. We quote Paul Hunziker who arrived at an interesting conclusion (1964): 'From the sum of observations one can conclude that the formation of sunspots is not the subject of solar forces alone, but is determined by the whole planetary system in one mighty process.' The wealth of variables, of which we can give only an elementary indication without the exciting details, gives us the picture 'that the sun is indeed a variable star, and the eleven-year cycle does not affect only the sun, but the whole configuration of energy in the planetary system.' Hunziker uses the expression 'sun' still in the traditional way; for our own use of language it would be superfluous to say 'not only the sun, but the whole . . . planetary system', because, as we have explained, they are synonymous.

The varying increase and decrease in the figures shows that the double rhythm of twenty-two years is only roughly correct.

Indeed, there is no fixed period; it varies between eight and fifteen years.

These irregularities, covering long periods of time, are today fairly accurately known. New sources of information have yielded data of an ancient time, long before direct observation was recorded. There is a natural world memory, that gives us carefully preserved information on these rhythms, and it works in the following way. Understandably, the tremendous events we have mentioned, the waves which permeate our whole planetary system, have an effect on the weather of the earth. This influences plant growth, which reacts to humidity and drought. These changing conditions leave a permanent trace in the varying thickness in the annual tree rings. Those who can decipher this script gaining an insight into climatic conditions, are called dendrochronologists. But that is not all. Even in older geological epochs the changes in weather conditions mirror the fluctuating life of the sun. Like tree rings, the fine layers of mud and other sediments in areas flooded during the ice-age demonstrate the ebb and flow of life's pulse in the sun's sphere.

There is a statistical procedure which elucidates the curves on the graph, so that the underlying tendencies of the many individual oscillations become apparent. Just as each instrument in an orchestra may have its own melody and volume, yet the ear hears one harmonious whole, so the working together of many natural events presents a unified whole. The so-called 'harmonic analysis' is a method for reading the music of individual instruments in a full orchestra. The graphs gained from the 'world memory' in tree rings and in geological strata represent the curves of the sun rhythm over vast periods of time. we can submit this curve to a harmonic analysis which shows that it can be understood either as a 'sum total' or as 'a result of many short, persistent periods, that run through the natural events without changing phases.' (Hunziker, 1964). This may be the reason why the search for the definite rhythm of sunspots had so little success and why the figure

3.3 QUI TOLLIT PECCATUM MUNDI

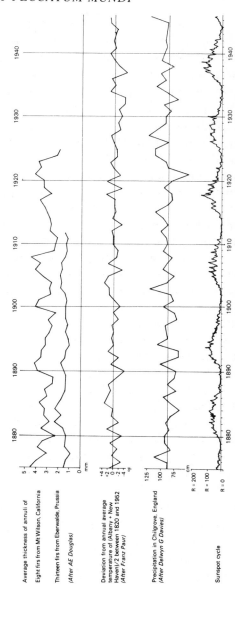

Figure 36. Dendrochronological curves. Variations in tree growth in USA and Central Europe are shown over a period of half a century and compared with temperature, rainfall and sunspot cycles over the same period.

of eleven which had been found later on possessed little strength of persuasion. The law of return is obviously only the result of many different rhythms that work together. Hunziker writes, 'The consistency and regularity with which this rhythmical impulse occurs through the centuries and millennia in the processes of nature, lead to a search for the pulsating factor in the field around the sun.' That means, in our own language: The processes that appear in sunspots or in the weather on earth have to be understood as life processes of the entire sun system. Indeed, 'the calculated rhythms occur in all constellations of the planets'.

Modern research confirms here the picture we drew of the planetary organism. The orbital oscillations of individual planets, their conjunction, opposition or quadrature, all this interplay makes up the healthy or unhealthy life of the parent sun just as the co-operation or disharmony among the organs determines the health of the human body. The skin reveals disturbance and destruction as long as there is friction and disharmony. A wide field for research opens up for the future, yet we have already acquired a general impression of what is to be expected.

Who then will not gaze with consternation at the mysterious being, the parent sun, who carries and surrounds us; who, in the joyous vitality of the young and in the wars of extermination, experiences and suffers the destiny of its children, the planets and all the beings dwelling upon them? Divine generosity, world-wide patience must be the quality of a being who makes the burden of its subjects its own burden. Deficiencies, weaknesses and deviations are accepted, and the shining face darkens until the intolerable effects on life as a whole — on the system — have been digested.

Must we not recall the words of John the Baptist, telling his disciples of a being who takes the burden of the world, the weight of cosmic sin, upon himself? *Qui tollit peccatum mundi.* From this aspect there is a harmony, a spritual-physical similarity of form; 'Behold, the Lamb of God, who takes away the sin of the world!'

3.3 QUI TOLLIT PECCATUM MUNDI

(John 1:29) It is the Lamb who embodies the sacrifice of God and lifts human opacity into the light of his radiant gold-shimmering, world-embracing fleece. It is the deed of the sun: the cosmic bearer of order entered into earthly humanity as Christ Jesus.

> Behold the Lamb of God.
> He will die for our sin
> On the tree of the cross
> To proclaim to all nations:
> God takes their sickness on himself,
> That mankind may know henceforth
> Not only their own small ego
> Is bearer of its darkness.
> (Morgenstern, 'Der Täufer')

Confronting this being, who would not hear the call to responsibility for his own actions? What if the deeds of the planets determine the radiance of the central star? What if into the hands of mankind has been given the mighty leader of our present time, the Messiah in all his suffering?

3.4 Who is the sun-king?

Again, a contrasting example may clarify what is meant. In the course of history, there have been many very different 'sun-kings'. We can think of the Natchez Indians of the Mississippi. They called their king the sun. His majesty was so elevated that his feet were not allowed to touch the ground in case they should be soiled and he was always carried in a litter. All great pre-Christian rulers calling themselves 'Sons of the Sun', earthly representatives of the heavenly god, were initiated sons of the mysteries. Whether they are of ancient history, primordial demigods of the Far East or the American continent, or supermen before the time of Christ — or even carrying the traditions through the generations down to our modern age — all have walked the inner sun path and brought super-personal qualities of rulership into their lives.

Today, we are inclined to reduce ancient truth to the plane of our own limited experience and talk about earlier times as if their peoples only imagined genuine relationships with their gods. Such opinions are a sign of cowardice in face of the facts. It is understandable if we fight with inadequate weapons. Any study of sun kingship in the past ought to lead us to profound meditation. We search the deepest region of the soul of mankind and look for long-lost faculties of the mind. Does the custom of the Natchez people not characterize a state of reality in which pre Christian people found themselves? The offspring of the sun had to be

3.4 WHO IS THE SUN-KING?

carried above the ground, lifted up as far as humanly possible above the gravity of the earth. In his very being, he belongs to a different world. If, in our meditation on history we discern something of the genuine cosmic divine relationship which primitive man possessed, our religions and scientific findings, our Logos who is Christ and sun, can converse with historical research.

However, leaving this mysterious realm of super-history, a case closer to our time, that of the French king, Louis XIV (1638–1715), demands careful understanding. He called himself *le Roi Soleil*, the Sun King. What did that mean at the beginning of the eighteenth century? Was he really a sunlike ruler? It appears that Louis XIV took this self-appointed title very seriously and regarded himself as the shining centre of his people and the centre of his country. He raised unconstitutional monarchy to its highest point by concentrating all power in his own hands. We look back uncomfortably to that period of absolutism which he personified. He wrote on his banner *L'état c'est moi*, (I am the state). All power is concentrated in one single despotic centre. Was this a 'sun kingship?'

Those who look at the sun as a centre of gravity and radiation, under whose rulership the whole planetary universe is driven towards life, may answer that it was; that sun, into whose 'absolute' sovereignty all other members of the system are delivered up, would be a shining example of the tyrannical monarch, and educated people with any scientific background may well be able to justify the self-chosen title of Louis XIV.

However, we experience a very great difference in the social sphere when we compare the vitalizing power of the sun with the tyranny of the absolute monarch of France. In truth, no bridge leads from the one to the other and no comparison is valid. On the contrary, the power which suppresses the free creative life of individual citizens, and paralyses and kills in the long run, would lead to a picture of a counter-sun if transferred into cosmic dimensions.

THE SUN

What would happen if the manner in which the sun orders, enlivens and works, as we experience it today, were transferred to human society? How would the organization of such a 'sunlike' community look? There is not yet a school or college where such a subject could be taught. We have prepared the way, in earlier chapters, for a blueprint for such a 'sun sociology' in its embryonic stage.

A story about Louis XIV tells us that he conversed with the writer Fontenelle one day. He summed up his life experience in the words: 'I must confess that my faith in honest men has diminished.' Fontenelle answered: 'There are many honest men, but — they do not seek out the king.'

This hit the nail on the head. Decent people, those who love truth and freedom find no room in the presence of kings like this. Only bowing and scraping courtiers congregate round them. This is an adjunct of absolute monarchy. A power-seeking monarch finds himself separated from the broad stream of his people and ends finally in isolation which can only be compensated for by brutal acts of surveillance and suppression. The noble element escapes because it feels threatened, or it gathers up its strength to destroy the system. Historically speaking, this is the best solution because it accelerates a necessary evolution. The age of absolutism gathered the tinder to ignite the French Revolution.

The effect of this kind of sun kingship is to disperse rather than to gather together and unite. No lasting community has ever been achieved; the system has dissolved from within. It has proved to be the opposite of a sunlike order.

Quite a different story has been told of Frederick the Great (1712–1786). Though he fails as a sun-radiant ruler, he does deserve credit for giving a completely new basis for the state. A hundred years had passed since *l'état c'est moi* was spoken when Frederick declared: 'I am the first servant of the state.' This great king fought many inner battles with his inborn Prussian attitudes and his own choleric nature, but this weighs lightly against the

3.4 WHO IS THE SUN-KING?

feat of recognizing the right path towards a more wholesome state. No-one changes his nature overnight. World history can afford to be generous in giving credit for insight and good intentions as against inadequate fulfilment. All his offences against his self-imposed new rule cannot hide the fact that he cleared the path towards organizing human society in a prosperous and living way by reflecting the strength with which the sun holds the planetary system together.

Countless stories about Old Fritz were circulated, showing his social skill in dealing with people, and the rough comradeship his subjects felt towards him. But one can also see how difficult it was for him to achieve the right relationship.

Shortly before the first performance of an opera the king ordered the time of the dress rehearsal to be brought forward because he would not be able to attend the première. The rehearsal began. After a short time, the king asked for the score (he was a musician and crossed out several passages. Then he ordered the conductor to change the performance, according to his alterations.

'That is impossible,' answered the conductor, 'the performance will be the day after tomorrow. Besides, I do have another weighty reason against the changes. I can only reveal this to his Majesty, when his Majesty is in a more gracious mood.'

Frederick listened with astonishment and remarked, 'I am not ungracious towards you, but towards the composer. Do state your reason.'

'Well then, I am the king in this realm!' said the conductor and he took the score back.

Frederick laughed and said, 'You are right! It will remain as it was.'

There is another story about a ball in the Potsdam Castle. Before the festivities began, the ladies quarrelled among themselves about the order of rank for the procession. Because the master of ceremonies could not find a solution, he asked the king for his advice. Old Fritz answered promptly, 'Tell the ladies, the stupidest walks in first!'

In its own rough way this Solomonic judgment echoes Christ's words: '. . . whoever would be first among you must be slave of all' and taken with Frederick's ideal of kingship, his hard answer evokes the same meaning and points to a sunlike social order. Did we not recongize the sun as a guiding star that submits itself to its creatures, and Christ as the spirit in whom the soul of the sun can incarnate in earthly presence? 'You know that those who are supposed to rule over the Gentiles lord it over them, and their great men exercise authority over them. But it shall not be so among you; but whoever would be great among you must be your servant, and whoever would be first among you must be slave of all. For the Son of man also came not to be served but to serve, and to give his life as a ransom for many.' (Mark 10:42–45). We are justified in acknowledging Frederick the Great as having made the first step towards being a modern sun-king.

3.5 New discoveries in the social sphere

Can one detect in political history how the sun spirit enters earthly conditions?

It would be rewarding to examine the modern development and transformation of states from this point of view. Do Christian attitudes gradually work their way into human social behaviour on the cosmic dimension? Research on this question will demand uncommon effort but promises uncommon results. It would gather different disciplines together and create new connections: History, politics and theology would have to unite their findings; social and natural sciences would meet in the same subject.

Combined efforts would show in a series of decisive events and outstanding personalities, how the archetypal Christ-impulse is gradually spreading throughout mankind, and however diluted, is beginning to work in many diverse ways. The two thousand years of Christian world history suffice to make certain stages visible. We noted just a few in the last section; a more thorough description must follow.

We turn our thoughts back to the Natchez Indians. Their governmental structure is a mixture of old and new forms. In the image of the sun-king who never touches the ground, we find a pre-Christian reality expressed, where the sun-god is worshipped at a distance, in extra-terrestrial spheres. However, at the same time, a form of human society develops around the sun ruler which is a prelude to later European developments. Kingship was not inherited. High office was earned through special qualities. The ruler was both son of the sun and king through the will of the people. Oliver La Farge says (1956, 28):

> Reports of this democracy were carried to Europe, especially by the French. There the idea of it was further elaborated, along with the philosophical idea of child of nature, the Noble Red Man. The influence of the first observations on the Indians can be traced through the French philosophers to the English, and with it went the concept of 'government by consent of the governed.' This idea was first put into specific words by the British philosopher Locke. Our own Thomas Jefferson, in turn, borrowed Locke's phrasing and used it, slightly changed, in the Declaration of Independence. The evidence is strong that the development of democracy in both Europe and America was affected by the sixteenth- and seventeenth-century contacts with American Indians.

This example shows how the new spirit began to spread over the earth. Around the archaic figure of the god-king there grew a structure of human power-sharing which expressed the situation in a changing world. This happened outside the Christian Churches, indeed it influenced European culture. We witness a process that illumines the cosmopolitan extent of Christ's deed. The spirit sun put its foot on the earth. Gradually a light streams from this event into the social landscape.

Let us return to Europe where the modern process of social evolution grew towards maturity.

During the time when Louis XIV occupied the throne, a movement was begun which was eventually to overthrow absolute monarchy. Charles-Louis de Secondat (Montesquieu), 1689–1755, who linked his ideas with those of John Locke, evolved a theory of power-sharing. He pointed out that it leads inevitably to disaster when all executive power is concentrated in one hand. Such accumulation of power has an innate tendency, indepen-

3.5 NEW DISCOVERIES IN THE SOCIAL SPHERE

dently of the attitude of the ruler, to oppress its subjects. This tendency to violate the dignity of men can only be counteracted if power is divided, reasons Secondat, so that the ruler only wields part of it. Another part can be given to representatives of the people in a parliament, another can be shared by a judicial court. Secondat became the spiritual father of the French revolution with his doctrine of power-sharing. Its threefold slogan: *Liberté, egalité, fraternité*, draws up a state system where life is ordered in a humane way, clearly distinguishing three realms. The triad, liberty, equality, fraternity had at the beginning more of a moral claim, but it became apparent later how convincingly these words denote three spheres of life which can be ordered independently according to the three passwords. If freedom is granted to everything that has to do with education and culture, equality can be experienced in the political realm as a self-understood human form of contact; and in the exchange of goods, it will be possible to arrange plans where human needs are met in brotherliness. Although these forms of realization of a threefold social order surpassed the capacity of the historic French Revolution, they lie hidden in the spirit of their origin like a comforting prophecy.

The most definitite progress towards the new is expressed in the two sentences, '*l'état c'est moi*' and 'I am the first servant of the state'. However, it needed much hard thinking to bring this fundamental swing into general consciousness and to make it fruitful for the future. An effort of lasting importance was made by Wilhelm von Humboldt, 1767–1835. He himself was a Prussian minister of state and the business of governing earned him his daily bread. Therefore, the title of his learned paper can be called sensational: 'Ideas towards an attempt to define the limits of the activity of a state.' This is a clarification of the real task of state government and of the framework in which the rulers have to operate if they don't want to overshoot the mark.

Humboldt seeks in his profound study to find a truth and one needs time and sagacity to follow him. An artistic temperament speaks directly and passionately and with a brief word touches the string that is taut in every man's breast. Friedrich Hölderlin (1770–1843) wrote: 'In the end, this truth will hold: the less a man experiences and knows about the state, whatever its form, the freer he is.'

Rudolf Steiner (1861–1925) puts the same matter in the shape of a question, 'What should the state abstain from doing for the well-being of mankind?' In this way a matter-of-fact treatment is possible. The question approaches the problem from the negative side like those previously mentioned. Perhaps it was historically necessary to begin by dismantling the old form. The crust that had accumulated on the monarchical form of government was too thick to allow any transformation to take place, before the shell was broken and the rubble removed. Rudolf Steiner was not content with a merely negative view. In his book, *Towards Social Renewal*, he contributed detailed positive plans for a new order of social life. In his many lectures, he enlarged on these ideas in the years after the First World War. He differentiated between different spheres of influence in the social organism. What Secondat had seen as necessary for the future was now put into modern form and worked out in practical detail.

The reason why this development, which aims at new forms of community life, has not been realized lies in the way that other forces have taken hold of power, preventing with brutal energy the transformation of states into human communities. The time for the tyranny of kings had passed, but in its place came on the one hand an absolutism of industrialized economy, and on the other the dictatorship of parties in rigid socialist systems. We live in a time of twofold compulsions used by state organizations to suppress us.

This is the *present* state of affairs. But the inner path leads mankind irresistibly towards new horizons. Victory for the light of the sun can be slowed down but never prevented. The cosmic

3.5 NEW DISCOVERIES IN THE SOCIAL SPHERE

hour which we may call 'sun on earth' belongs to the future, also in the social sphere.

In the midst of chaos between ruling powers, some seeds of future events can be detected. However little the rulers of democratic countries may understand the forces of the future, they do realize more and more clearly that effective government has to take notice of people's moods and opinions. The parliamentary representation of a nation in modern party politics proves a misleading solution. Government depends on the involvement of every individual. That is the reason why opinion polls are valued so highly today. It is a pointer in the right direction but proper conclusions for future forms of government have not yet been drawn. Polls are used in order to discover popular opinion and to catch votes, but important decisions are still taken by a small majority, sometimes bought over the heads of the public by a ruthless lobby. But a grain of hope is contained in the effort of opinion pollsters. Whoever is in the centre, be it the government, a council, or a board of directors, they have to be in the picture concerning everything that happens within the system right down to the last detail. (The tools for such gigantic service on state level are to be found in the mass media and the computer industry.) So far we have dealt with the political sphere.

In economic life, we can find small beginnings of new forms of community. While the majority of those involved get enmeshed in a jungle of competitive decision making, a few firms show examples of genuine co-operation which satisfies all members and brings successful financial returns. We are thinking of the original system by Spindler (1964) who among others deserves to be known in wider circles. Without consciously being aware of the connection he established an ordered community within his firm where the social sun-order is at least partially brought into reality.

3.6 The hierarchy as an order

One can learn something from Spindler. Any special benefits bestowed by a director on his workers are taken right out of the sphere of charity and are regarded as human rights ensuring a decent treatment of everybody. This seems exemplary. It is worth noticing the importance he allows to the plenary meeting of all workers. He gives surprising examples of the results frankness and mutual trust can bring. A community-spirit is carefully nurtured, a right to share in the responsibilities is coupled with the right to voice opinions. Those who feel valued as co-workers are prepared to exert themselves to the utmost. Spindler wrote (1964, 342):

> In 1955 a new law decreed full equality for men and women in the textile industry. This would normally be achieved by raising the wages for women to the level of male workers. In this particular case the increase would have been 24 per cent for women. After thorough discussions within the fully informed committee and plenary sessions, a formula was worked out together with the directors that the men gave up 2 or 8 per cent of their pay (depending on differentials) in order to reach a new common basic pay for the two groups. In a secret ballot which followed these deliberations, 80 per cent of one group in the work force voted to forgo 8 per cent and 90 per cent of the second group voted to forgo 2 per cent.
>
> . . . The year 1958 was difficult for our firm for two reasons. The previous year had seen great changes in pro-

3.6 THE HIERARCHY AS AN ORDER

duction, which meant rationalizing and contracting. The cost of this programme had not yet been absorbed when the new year brought increasingly harsh competition on the textile market.

On April 1, 1958 a rise in wages and salaries came into force which constituted an extraordinarily heavy burden. Again we provided full information and discussed all questions in committees and with the directors. The problem was whether some important investments could be carried out or not. The long experience in decision-sharing together with the possibility of checking the figures individually, convinced the workers that the proposed investments were right and necessary. They decided to forgo a rise in wages for nine months that had been nationally agreed upon by employers and trade unions. For this firm, the rise came into force on January 1, 1959 without back payment.

As Spindler himself insists, these examples have to be treated with care. They must not be misunderstood as clever tricks to increase production by surreptitiously obtaining sacrifices. Should the partnership be misused, such dishonesty would reveal itself very quickly and lead to failure. The examples should only prove that a true co-operation includes the sharing of common risk and failure. Partnership has to be understood in a similar way to a team on a rope, tied together out of free inner decision in order to climb a certain mountain peak. While such a team hangs on a rock face nobody can go his own way without endangering everybody else.

Spindler's examples, chosen with delicate understanding, are a first-class aid for anyone wishing to inform himself about modern community building. One can see how relationships consisting of a tangle of ambition, suppression, greed, enmity and passion may be lifted almost overnight from the level of power struggle into the light of a future where the dignity of man holds sway, simply through different treatment. One can recognize the first steps of a 'young path' as Spindler calls it, a path which we, who are searching for a sociology of the sun or for a Christian sociological science must keep in sight.

It remains to ask how the tasks within a partnership can be fulfilled without violating fundamental laws of co-operation. How can equality in rights and duties be reconciled with the diversity of job distribution? This is a decisive question for any small or large group within society. Out of different demands posed by any task a gradation of jobs and responsibility arises. There simply has to be somebody like a foreman, a manager and a director in business, or bishops, archbishops and cardinals within the Church hierarchy. How can the gradation in spheres of influence be planned in such a way that there is no patronizing and no rulership of men over men?

The answer is as simple as it is revolutionary. By not projecting any 'above and below' into the hierarchical structure, but rather allowing different spheres of responsibility to intersect, penetrate and embrace each other. What does this mean in practical life?

Spindler shows how it can be realized in industry (1964, 224). 'In the business which I manage, every worker has the right to receive an explanation for every order he is given.' He who has such good insight that he is able to explain it to others and to awaken an understanding for what needs to be done, he is the person to give orders. 'This demands a new type of superior.' The old method of issuing orders was in most cases identical with the holding back of insight into the complexity of the assignment. This method is now obsolete and no longer achieves anything. Henceforth the importance of above and below ceases to exist.

The new kind of partnership is far more uncomfortable for both sides than the old style of rulership. A worker can no longer play the role of the reluctant or indifferent yes-man; and the boss now has the task of mediating understanding. His orders have to possess such transparency that the will-power in the worker is called upon. Someone expressed this with the words, 'the boss

3.6 THE HIERARCHY AS AN ORDER

has to become the helper of the subordinate.' This echoes the ideas of Frederick the Great and builds a bridge from the young path of the modern businessman to the new forces in the political realm.

The true task of those in responsibility is to be a helper. He has to make it possible for workers in a group to let their individual efforts stream together into a meaningful and orderly whole. This kind of activity of a superior can be compared with the spreading of light, a vocation of light. Whoever wishes to carry such responsibility must be able to survey the whole complexity, and communicate his vision to those others who cannot acquire it from their own standpoint. He needs an intensified perception and the gift of making it clear to others to ensure an ordered movement of the whole. (The old word *episcopus* meaning bishop — the one who surveys, or the overseer — is well chosen. It expresses the first part of this task while the modern word leader emphasizes the second task.)

On this young path it is out of the question that anyone can gain promotion simply on the length of his service or through graft. Any mistake in the filling of promoted posts will reveal itself swiftly. The question of qualification for a particular position becomes one of ability. An example from the African campaigns in the Second World War may give us a picture: 'Captured British battle dress was frequently worn by German officers. It was so difficult to attach military stripes or badges that they were worn without any distinguishing marks. No problem arose. The others soon noticed who had something to say, though he wore no shoulder stripes or braid.' (Spindler 1964, 339).

Whatever name is given to the boss, the 'young path' has power to reveal qualities of leadership, the quality of recognition and an eye for talent. The highest position can only be filled 'by a personality invulnerable through the fairness of his behaviour. (Spindler 1964, 311).

The word fairness stems from the word fair — blond. The

vocation of light seems transformed from clarity in practical realms to a moral quality of justice, a unifying radiance of the soul. This is what we expect today of a leader.

Spindler describes the manager of tomorrow (1964, 325):
> His task as a trustee is fulfilled by the manager through a co-ordination of all forces working in the human sphere and all material factors within the working community. Everybody takes part and feels engaged within a responsible whole.
>
> ... In order to act as a trustee, of all persons concerned, the manager has to attain a spiritual point of view which lies above any partisan interest.

The trustee expresses an acting out of a 'trust which others have in the trustee. It expresses a profound character and it reminds us of the hands of the sun with which the leading planet carries the system of the universe. The sun serves the harmony and well-being of the whole world.

Could we not transform our understanding of the heavenly hierarchies from our experience gained in human life? The word hierarchy means 'holy order' and was used for the heavens before it was transferred to human organization on earth. Can we still understand the order of angels in the picture of a ladder? Should we not transform this picture completely to find the truth with respect to our advanced consciousness? If we could raise this question afresh and find answers for today as a signpost to reality, then we would find a new relationship to heavenly truth and illumination for earthly society; a fruitful meeting between theology and sociology.

Indeed, we can see the hierarchical order of angels in such a way that the succession is expressed in magnification. Each successive angelic task is characterized by wider vision and greater facility in communication. A higher spiritual realm proves itself by greater strength and wider radiance until the godhead in its all-embracing

3.6 THE HIERARCHY AS AN ORDER

Figure 37. Social organization based on the sun germ cell. Whitsuntide. Westphalian master, c. 1380.

being meets our minds in the further ground of heaven and earth, in spiritual and physical world totality.

We have the hierarchy before us in successive revelations as an order of planetary competence just as in the picture of the solar system the order in space makes the omnipotence of the sun understandable to us.

3.7 'My kingdom is not of this world'

In the near future, people may perhaps look back to our age and shake their heads in pity when contemplating the scientific superstitions that held a large part of mankind spellbound. The sun is thought of as a gaseous sphere holding its planetary system together by gravitation which means that the sun actually forces the planets to follow in eternally repeated circles round its 'centre of mass' with the same force that makes a stone fall down on earth.

Today, new insights promise to dissolve the laws of Copernican and Newtonian astronomy that had been considered irrefutable. An adviser to NASA and the US Air Force, a man with several degrees and an expert in these fields, said to the author 'One would not like to admit what observation has shown in the last years, namely that the sun is not a body at all, as previously understood, but consists of plasma.' (The term plasma means a non-material state.) Wise heads are gradually discovering that previous interpretations of solar research led to error. The expression 'plasma' is used for a gas consisting of an approximately equal number of positive ions and free negative electrons. This sounds too evasive; fundamental rethinking is necessary for a true understanding of the sun, but this is avoided. Yet to speak of a 'plasma-sphere' and use ideas far removed from ordinary earthly phenomena leads close to a non-material position and it then becomes necessary to examine old theories that this kind of sun

3.7 'MY KINGDOM IS NOT OF THIS WORLD'

'attracts' by gravitation. This is a compelling demand. We don't need it for our own interpretation of the sun because we started from different premises. Nevertheless, we welcome this attempt to overcome old ways of thinking about the sun.

We never looked at the sun as a 'ball' of any kind but rather as the macrocosmic life-body in which the planets move as inner organs. Was it not misleading to speak about the attraction of masses? A living unity does not need gravity to keep its parts together. At another level the polarity of sun and earth became clear to us. Opposites are bound together because they are polarities. Only together can they form a unit. They do not need to borrow qualities from each other. Just as little as a husband is fettered to a wife because she acquires masculine qualities, just as little are sun and earth fettered together by the sun having explicitly earthly qualities. On the contrary, the very differences in the heavenly bodies links one to the other as male to female sex, or as the magnetic poles reach out to each other and hold together. Another pointer to the kind of adhesion in the solar system was given in the picture of the cosmic whirlpool which unfolds its specific action out of a hollow, sucking centre.

This problem appears again on a different level from the social viewpoint we applied in the last chapter. True to our principle, we want to relate the question to practical experience, to build up a living science. What does our own life tell us? How do human communities keep together? What makes a personality attractive so that others rally round him? Would it not contradict our experience if somebody claimed despotic exercise of power as the uniting element, keeping members of a family, club or society together most effectively? The opposite is the case, as we know: Today, any trace of authoritarian behaviour by government officials is decried. In the present age, tyrannical government provokes nothing but rebellion and finally revolution. Every absolutism works towards dispersal, never towards gathering. It was different in earlier times. We agree today that forms of government which enslave the populace are outdated. Where does this new direction come from?

The answer in brief: the evolution of humanity stems from a new spirit which has entered earthly conditions since Christ walked on earth, the sun spirit with its inherent peaceful guiding power. After an initial two thousand years, the presence of this spirit shines more and more clearly in modern times.

Let us return to the question: which kind of behaviour gathers other people round a personality and forms well-ordered community? Why do we like to join another person, when do we feel his orders, not as a punishment, but rather as suggestions which we follow voluntarily? Sometimes we meet a person in whose presence we have a feeling of well-being, we wish to stay close to him and always like to return to him, but we don't know why. It may be an old mother or a talented young man, a friendly family doctor, the boss in the office, the cook, or the girl at the reception desk . . . independently of age or profession, some stand out as being attractive. We can discover what they have in common. The mother, doctor or receptionist were there for anyone who happened to meet them. They listened to us, remembered what was the matter last time, they ask further questions on how we coped with examinations, a sick child, with our itinerary for travel, and so on. Quite simple things, but the one who is being asked does know that a soul opens itself for him. He feels welcomed, affirmed, encouraged and accepted. Touched by a loving radiance.

Two things coincide here. A being is truly recognized and shone upon by a genuine light from the heart. This twofoldness brings the attraction about. Sometimes we meet a teacher who cultivates a relationship with his pupils and shares their destiny far beyond the school. Pupils feel attracted to such teachers, they revere true leadership in him. (If a teacher knows the souls entrusted to his care so well that he anticipates their needs, then he has solved the task of education perfectly. There will be no problem of

3.7 'MY KINGDOM IS NOT OF THIS WORLD'

discipline nor a generation gap. Sensitive understanding of human nature coupled with warm-hearted openness for every single pupil constitutes the key to all education.)

'I know my own' (John 10:14) is the attitude which Christ has to his disciples and which makes Christians rally round him and remain united with him in this 'knowing'. In it we see the origin and the aim of what individual people achieve by loving openness towards others. In this way they gather a circle of workers, lovers or friends around them. It is this power that enables some members of staff to carry superior tasks as we have heard from Spindler's 'young path', the power which a doctor develops, caring for the sick, or an active teacher. It is the power which brings a priest into a free and intimate relationship with those who trust him.

We find the opposite characteristics in Louis XIV. People who knew their worth kept away from him, nobody could honestly feel that he 'knew' them or understood them. Somebody with an understanding of his own worth would have felt unappreciated or misunderstood. This ruler did not emanate an interest in his fellow men. How could he turn his attention to them when he himself was the state, he alone and nobody else? He had only one intention, to suppress his subjects with unyielding power. He did not work with ability and discernment but with wilful impetuosity. He forced his system of earthly tyranny, using weight against human lives. The ruler used forces of the earthly realm alone.

'I know my own' is thrown into relief. The genuine perception of the other person reveals itself as a sun-born quality of light. When the other person is known as a 'thou', all social achievements of compassion, of sharing in joy or in knowledge, every form of love is transformed sun-power brought into the human realm. A social order that builds upon the recognition of the other being is a light-filled order that does not originate in the earthly sphere of gravity, it belongs to another world. No wonder that the bearer of the sun-majesty, the Christ, says, 'My kingship is *not* of *this* world' (John 18:36).

THE SUN

Once more we see the sun before us in a moving vision. Everything we have gathered about its being culminates in the postulation of this star as a gigantic moving ocean of light. But its waves do not break in a wild or disorderly fashion. We discover large-scale rhythms in its tidal floods. Like a kind of breathing and a circulation of life-blood stretching out from the centre to the furthest member, gathering it to its world-wide heart, thus the sun organism pulsates untiringly. We see the corona stretch out and the same is observed by astronomers in the zodiac at night. The radiant body transposes its sentient body outwards, living with and suffering with the planets. The hands of the sun touch each part of the far-flung living sheath, as gently as the tender hands of a healer to whom even the slightest disturbance is revealed. When all is embraced the processes of contraction follow. We observe how the corona shrinks and the pyramid of light of the zodiac is reduced. The light body contracts towards the centre. Everything that had been experienced, all damage that has to be healed, all pain is unloaded there and has to be digested, smelted into counter-space, metamorphosed into new radiance, causing a brief darkening of the shining organ. This is a time of many large sunspots, the so-called maximum. Again, expansion follows at regular intervals when conditions of life in the system warrant it; while the sun is active in a wide radiance the disc remains free from sunspots, the time of the minimum.

The two stone reliefs on St Jacob's Church in Tübingen may symbolize this expansion and contraction of the sun-hands. Do they express a contracting and widening as described in the second part of this book? Amazing treasure! These apparently primitive runes reveal in their threefoldness the secret of the sun sphere.

Figure 38. The three sun stones from the walls of St Jacob's Church in Tübingen. If, as seems probable, the three relief stones, now in separate places, originally belonged together, then we have here a picture of a dynamic unity.

3.7 'MY KINGDOM IS NOT OF THIS WORLD'

Incessantly as breathing or heartbeat, the living rhythm of our mother star takes place. Perceiving, digesting, radiating, this is the threefold accord of its loving devotion, the foundation of order in the kingdom of light. Perceiving, transforming, illuminating, it carries everyone who follows it and keeps them in health.

The words, 'My kingship is not of this world' which remain a riddle to theologians or have been misunderstood as a negative alienation of the world, suddenly reveal their actual meaning. Listening to these words of Christ in the light of this book, we can understand the secret of the sun's attraction. One day it will dawn on everybody: no force of this world, not the force of mass, not gravity, ties the planets to the central star, but the light and love of the cosmic shepherd whom the 'sheep' follow in free responsive love, recognizing him as their leader. Carried down from the realm of the majestic sun to the earth this word becomes the foundation stone of Christian social order. The ancient picture of the shepherd arises anew in Christ's speech, rejuvenated by the sun. He who 'knows his flock' and wishes to be the servant of those needy of help, is walking before them. Only to him can a freedom-loving man ally himself in love and with willingness to carry responsibility He alone can show the way into the future kingdom of light for all mankind.

3.8 Sun-god and brother of man

In the kingdom of light we saw the sun spirit's brotherly activity in the history of the earth.

This glimpse of the greatest secret of the world should satisfy a deep inner longing in the human soul. If it does not, it may still disturb some unquestioned prejudices. Before the turning point of time, people had knowledge of cosmic divinity but they knew that gods had to be worshipped from afar. Just as our eye is blinded by the direct glare of the sun so any immediate confrontation with the divine must hurt or even destroy the human soul. There remained the hope and promise that one day the *plērōma* or fullness of God would enter the earth in human form and unite mankind with its original spirit home, where it could not return on its own strength.

Many wonderful legends spoke prophetically of divine heroes visiting the earth in pre-Christian times.

It becomes understandable that those people who were open enough to recognize the Christ incarnated in Jesus interpreted the event with a fervour bordering on madness, emphasizing the earthly quality of his incarnation. They clung to the verse 'the Word became flesh'. In the beginning, there were hot disputes about the origin of the Son of God from beyond the earth and the reality of his bodily reincarnation. Victory was finally won by those who held a one-sided view of the nature of Jesus and had lost understanding for the cosmic spirituality of his being. Churches bound by dogma still speak of his divine origin and character, but many honest people find the words empty because they cease to connect astronomical concepts with words of the spirit. They are left with the 'simple man from Nazareth' who does not compare at all remarkably with other great men, like Socrates, Buddha, Confucius. Why should he be the unquestioned winner in such a contest?

The longing to see God incarnate and close to man can only be satisfied if the name of God is not empty, but filled with vivid life, with colour and content. The humanity of Jesus gains its incomparable depth and weight because it rises up as a sacrificial deed of the divine spirit from the cosmos.

We have realized that the victory over the adversary through Jesus Christ consists in the rejection of the temptation to act as a cosmic bearer of power. It initiated new paths where instead people could gain the transforming sun power in gradual evolution, free of all devastating heat, metamorphosed into a blessed form which we can understand. We experience the meeting with the Christ-spirit in no other way: where he teaches, where he heals, where he wakens the dead. He brings the 'light of the world' transformed into rays of love which liberate man in a health-giving and renewing way. On all his wanderings, in all his deeds, we experience how in Jesus the Christ enters the earthly world as a human being.

We must not forget any one of the three different members: not Jesus as the earthly bearer of the spirit; not Christ, the sun hero who descended from the cosmos; and not the insoluble link between both parts of his being which is expressed in the word 'as man' allowing our share in this unique event, for all future time. In this threefold aspect is contained the whole truth of sun and earth and man.

If we recognize 'the wholly other' in its cosmic greatness and find it again in the 'kingdom of light', then we have begun to

3.8 SUN-GOD AND BROTHER OF MAN

unravel the fundamental truth of world history with modern methods.

At the beginning of the nineteenth century a forgotten prophet shared the knowledge of Christ as the sun. For this historic deed alone Joseph von Auffenberg deserves to be known far more widely. Doldinger (1970) introduced this neglected personality to our time telling of their visit to the Brera in Milan in Auffenberg's own words:

> In the very first room I could not tear myself away from an excellent copy of Leonardo da Vinci's *Last Supper* while the sun was shining directly on the head of the Saviour from above. Friedrich nudged me several times, 'You won't be finished in a week if you dawdle like this.'
>
> I assured him that I would stay as long as the sunlight was there. He took snuff and said contentedly, 'Something similar was said to me by the Duke of P . . .'
>
> 'Go to the devil with your duke,' I begged, 'and leave me in peace to contemplate the Divinity, lit up by his own sun.'

May these sketches on sun research serve the recognition of Christ as 'the Divinity lit up his own sun'. May they encourage people to seek communion with him where he sits at table like a brother. It is the altar, and at the same time the sphere where world-wide meetings take place. This wish accompanies this book.

Epilogue: Limitations and gratitude

This book differs in two ways from the ordinary. It is neither scientific nor religious in the traditional manner. And so it may irritate the reader, astonish, disappoint, or anger him, depending on his temperament. The twofold title subtitle, 'Ancient mysteries and a new physics' may have aroused expectations that were fulfilled in an unexpected manner. Hope remains that no-one will put the book down without having learnt something new. If science and religion are to be redeemed from enmity or isolation and brought to a genuine meeting from which something new can be born for the world, they presuppose a genuine will to change. Imaginary safeguards have to be sacrificed, new horizons gained and rigid concepts made fluid.

Science can only achieve greatness if it shares the path with mankind. It should not lose courage before this task. Critical scientists complain that science has lost touch with the life of man. Lehrs wrote that science retreated to the point of view 'of a single colour-blind eye' (1951, 30). Walter Heitler writes (1963, 94f): '... Physics describes the causal and quantitive aspects of the world, a sort of image projected on the causal and quantitive plane, but certainly not a complete image.'

But science can do more. Centuries of training gave us skills which we can use willingly. We are able to find the path with mankind in our time. It is a question of discipline, of patience and inventiveness. No step must be taken without involving man with total attentiveness to the full stretch of his mental capacities. Many one-sided practices have to be given up, but important new possibilities appear.

To include man himself means to use human perception as a basis for research: to accept the world as it really appears to us. Let the phenomena speak. They express the truth contained within. Phenomenological method is also called 'Goetheanism' because Goethe used and described it. He can become a teacher for everybody who seeks the path of man in the study of nature.

If the scientist is involved as a complete human being, he himself will undergo a transformation. He will become involved in the object of his research. *Science leads to genuine objective involvement. Thus it reveals its inner religious strength.*

To live in the sun. For one who follows step by step, contemplation becomes experience. The results cannot leave us indifferent, because our personal activity is present in every phase. Living in the sun fundamentally changes our lives.

We wish to distinguish between our endeavour and that of others, for example Ditfurth's book, *Children of the Universe* (1975, 290), which contains the results of astronomic research in a fascinating survey. The value of the work should not be dismissed, the author deserves admiration. The construction of the cosmos is delicately balanced, no member could be left out without endangering the whole. Take the moon away and the building collapses. These words made us sit up; is the universe recognised, then, as an organism? Not at all: 'The vastness of space established the preconditions for life . . . Earth and sun, moon and solar system all helped to bring us into being . . . The cosmos . . . brought us forth and maintains our being . . . This fact should give us faith as we travel on our way — even if no one can tell us where the journey ends.'

That is the end result of gathering details and heaping them together. But the result alone is not the most important thing. Whoever examines the procedure will find the root of the evil

EPILOGUE: LIMITATIONS AND GRATITUDE

quickly. The thought clings to hundreds of hypotheses that remain alien to our perception and can only lead to frustration. No wonder a picture of the world emerges with which we cannot have an inner relationship as human beings.

We have consciously avoided having anything to do with this type of representation. The principle of immediate sense perception and the understanding which follows from this method give us a complete picture in which every part is closely connected, with man acting as the medium.

It is now time to thank the man who pioneered the Goethean path. He recognized in it a uniquely precise method and perfected it. Without the incentive of Rudolf Steiner, his numerous writings on solar research and Christology, this book would never have been written. His anthroposophical statements on questions regarding the sun, made under the most diverse circumstances, gave the impulse for the attempt to present a picture, based on astronomical observation, and drawing together a number of other sciences.

What must religion do in order to meet the challenge expected of it? As was the case with science, religion must reconsider its original purpose.

Religion cannot and never could have had any other meaning than that of interrelating the divine spiritual with earthly human life. The word *religio* may be translated as 'conscientious consideration, careful observation'. To walk before the countenance of God is the origin and goal of the activity, and between ancient and recent times it is only the direction of seeking the divine being that has changed.

Once it was natural that human beings saw their gods, and religious ritual consolidated that relationship, and helped later to hold and cultivate the waning power of perception. The whole inner life of the soul was thereby made backward-looking to the disappearing primordial light. The modern religious man must take the opposite way.

After the last glimmer of the old clairvoyance had passed, there came a turning-point in time. As God, descending from the sun, Christ came down to earth to open for man a new epoch of closeness to the divine world. He instituted the religious ritual of the breaking of bread and showed his disciples the way, through repetitive practice of union with him, to a future conscious perception of God, face to face. During the early Christian development from the central event of the new religion, there grew up other sacraments, related to the manifold human biography, opening the door to exact observation and inclusion of the spiritual world.

Practised in society, Christian 'consecration' ignites the consciousness of the spirit, directs us towards a future evolution of God's presence.

Religion guides towards genuine broadening of consciousness. It demonstrates thereby its innate scientific productiveness.

Here a reservation arises. On the road we have taken, we dissociate ourselves from all religious enquiry that believes it must be founded on textual evidence or dogma. If we ourselves refer to Holy Scripture or other traditions, it is but to establish the worldwide, timeless solidarity of this knowledge among all those seeking spiritual evidence. For us the Gospel is not to be found in a book, but is a living process in which we can participate in each completion of the new Christian ritual. Whoever goes on spiritual voyages of discovery is gladdened by every confirmation of his own findings, no matter where he may encounter it.

In this respect, a remarkable manuscript may be mentioned, that turned up as a friendly greeting upon the completion of this book. The *Redentin Easter Play* written in Middle Low German, is preceded by a Latin sermon that gives five comparisons of Christ with the sun. The mystery play contains the descent to hell and the resurrection, and is deeply connected with the esoteric aspect of Christ as the sun, though the sermon is written in a

EPILOGUE: LIMITATIONS AND GRATITUDE

traditional style. Perhaps other known or unknown passages in literature can be found where the Christ-sun is hinted at.

Rather than end with a backward glance to older sources we shall point to the well from which our knowledge springs. The lens through which we see spiritual facts, and find illumination of the inner configuration of the world is the continuous celebration of Christian ritual. The Act of Consecration of Man presents us with a wealth of experience which brings the cosmic picture of the sun together with the divine reality of the Christ. In the cult we walk with Christ and receive his creative life into our own innermost being.

> And now you wish to give thanks to him
> But your gaze is met by the sun,
> Flaming and burning sun. . . .
> And you stand in amazement, touched by love itself.
> (Winfred Paarman 'Die kommende Kirche')

Photographic acknowledgements

Hirmer Photo-Archiv, Munich, 1; Fraunhofer Institut, Freiburg, 2, 3; Max-Planck-Institut, Heidelberg (Hans Ruder) 4, (P. B. Hutchinson) 7; Deutsches Museum, Munich, 5; *Stern und Weltraum*, Bibliographisches Institut, Mannheim, 8, 25; Verlag Urachhaus, Stuttgart 9, 38; Staatsbibliothek Preussischer Kulturbesitz, Berlin, 10; Ullstein-Bilderdienst, Berlin, 12; Karl Hublow, 14, 15, 16, 17; Fratelli Allinari, Florence, 18, 19; USIS/NASA 20; Verlag Freies Geistesleben, Stuttgart, 26, 27, 29; Philosophisch-Anthroposophischer Verlag, Dornach, 31; Bildarchiv Foto Marburg, 32. Other photographs are the author's.

Bibliography

Adams, George & Olive Whicher. 1952. *The Plant Between Sun and Earth*. (Reissued 1980. London: Steiner.)
Adams, George. 1965. *Strahlende Weltgestaltung*. Dornach: Philosophisch-Anthroposophisch.
Bock, Emil. 1979. *Schwäbische Romanik*. Stuttgart: Urachhaus.
Boschke, Friedrich L. 1970. *Erde von anderen Sternen. Der Flug der Meteorite*. Frankfurt: Fischer.
Cloos, Hans. 1954. *Conversations with the Earth*. London: Routledge & Kegan Paul.
Ditfurth, Hoimar von. 1975. *Children of the Universe. The Tale of our Existence*. London: Allen & Unwin.
Doldinger, Friedrich. 1970. *Das Asyl*. Freiburg: private.
Dostoyevsky, Fyodor. 1982. *The Brothers Karamazov*. Harmondsworth: Penguin.
Edwards, Lawrence. 1982. *The Field of Form*. Edinburgh: Floris.
Grimm, Hermann. [1889]. *The Life of Raphael*. Paisley & London: Gardner.
Heitler, W. 1963. *Man and Science*. Edinburgh & London: Oliver & Boyd.
Heyer, Karl. 1956. *Das Wunder von Chartres*. Stuttgart: Mellinger.
Humboldt, Wilhelm von. 1962. *Ideen zu einem Versuch, die Grenzen der Wirksamkeit des Staats zu bestimmen*. Stuttgart: Freies Geistesleben.
Hunziker, Paul G. 1964. Essay in *Elemente der Naturwissenschaft*. Nos. 1 & 2. Dornach: Philosophisch-Anthroposophisch.
Jeremias, Alfred. [n.d.] *Ausserbiblische Erlösererwartung*.
——1911. *The Old Testament in the Light of the Ancient East*. London: Williams & Norgate. New York: Putnams.
Jung, Carl Gustaf. 1953. *The Psychology of the Unconscious*. (Vol. 7 of *Collected Works*.)

———1956. *Symbols of Transformation.* (Vol. 5 of *Collected Works.*)
———1957. *Collected Works.* London: Routledge. New York: Pantheon.
Kiepenheur, Karl Otto. 1957. *Die Sonne.* Berlin: Springer.
———[1959]. *The Sun.* Ann Arbor: University of Michigan.
Kühn, Rudolf. 1962. *Die Himmel erzählen. Astronomie heute.* München, Zürich: Droemersche.
La Farge, Oliver. 1956. *A Pictorial History of the American Indian.* London: Spring.
Lehrs, Ernst. 1951. *Man or Matter.* London: Faber.
Locher-Ernst, Louis. 1937. *Urphänomene der Geometrie.* Zurich: Orell Füssli.
———1940. *Projektive Geometrie.* Zurich: Orell Füssli.
———1938. *Geometrisieren im Bereiche wichtigster Kurvenformen.* Zurich: Orell Füssli.
Lusseyran, Jacques. 1963. *And there was Light.* (Reissued 1985. Edinburgh: Floris.)
Meyer, M. Wilhelm. 1897. *Das Weltgebäude.*
Neupert, Dr Werner. 1971. *OSO 7.*
Redentin Easter Play. 1941. New York: Colombia University.
Richter, Gottfried. 1965. *Chartres.* Stuttgart: Urachhaus.
Schmidt, Thomas. 'Poläritaten der Erscheinungen am Fixsternhimmel'. *Sternkalender 1968/69.* Dornach: Philosophisch-Anthroposophischer.
Schütze, Alfred. 1960. *Die Mithras-Mysterien und Urchristentum.* Stuttgart: Urachhaus.
Schwenk, Theodor. 1965. *Sensitive Chaos.* London: Steiner.
Sölle, Dorothee. 1968. *Phantasie und Gehorsam. Überlegungen zu einer künftigen christlichen Ethik.* Stuttgart: Kreuz.
Spindler, Gert P. 1964. *Neue Antworten in sozialen Raum.* Düsseldorf, Wien: Econ.
Steiner, Rudolf. 1972. *The Social Future.* New York: Anthroposophic.
———1977. *Towards Social Renewal.* London: Steiner.
Streit, Jakob. 1984. *Sun and Cross.* Edinburgh: Floris.
Stumpff, Karl. 1957. *Astronomie.* Frankfurt: Fischer.
Whicher, Olive. 1970. *Projective Geometry.* London: Steiner.
———*See also* Adams, George and Whicher.
Wolf, Rudolf. [n.d.]. *Geschichte der Astronomie.*

Index

Figures in *italics* refer to illustrations

absorption spectrum 161
Adams, George 138, 159
ankh (Egyptian sign)☥ 46
apostles 105
Areopagite, Dionysius the 86
Auffenburg, Joseph von 225

Baptism of Jesus 73, 94–97, *75*
Bride, Legend of St 81

Cassini, Domenico 174
Celtic Messianic expectation 81
Chillon Castle, Lake Geneva 177, *178*
Christian 106
Christian Community, The 180
collective unconscious 9
Copernicus, planetary system of *33*
corona 21–23, 31–34, 53, *20, 22, 32*
counter-space 134–38

dendrochronology 196, *197*
Dionysius the Areopagite 86
Dostoyevsky, Fyodor 91
dove (at Baptism) 95

Egyptian Messianic expectation 79

emission spectrum 161
Euclidean geometry 135
Extern Stones (Teutoburg Forest) 81, *80, 82*

Fabricius, Johannes 186f
Frederick the Great 202–4

Germanic Messianic expectation 81
gherkin experiment 154–56, *155*
Goethean science 229

hands of sun 44–49, *47*
Hölderlin, Friedrich 208
Humboldt, Wilhelm von 207
Hunziker, Paul 195–98

initiation 78, 105

Jacob's Church, St (Tübingen) 44, 220, *45, 221*
John the Baptist 73, 96, 198
Jung, Carl Gustaf 9–11, 95

Kiepenheuer, Prof. Karl Otto 63
Kühn, Rudolf 53

lemniscate 166–174, 177, *165, 167, 168*

Locher-Ernst, Louis 135f
Louis XIV 201
Lusseyran, Jacques 28, 107–9

man 119–22
Messianic expectation 78–81, 86f
Montesquieu (Charles-Louis de Secondat) 206f
moon 52

Natchez Indians 200, 205f

Paul, Apostle 48, 86
Persian Messianic expectation 86f
plant 119, 157–59
projective geometry 135–50

Raphael, *School of Athens* 86, 87
— *Transfiguration* 112, *111*
Redentin Easter Play 230

Scheiner, Christoph 188

Schwenk, Theodor 69
Secondat, Charles-Louis de (Montesquieu) 206f
Spectrum analysis 161, *162*
Spindler, Gert 209–14
Steiner, Rudolf 208, 229
sun king 200
sun's rotation 128f
sunspots 185–90, *186*, *191*, *195*
— period 188, 194

Temptation of Christ 99–101

vortex 68–72

Wells Cathedral, Somerset 177, *179*
Whicher, Olive 153, 159
whirlpool 68–72, *70*
Wolf, Rudolf 188, 194

zodiac light 35–41, *36*, *38*